Electronics IV

Electronics IV

A textbook covering the analogue (linear)
content of the Business and Technician
Education Council's scheme for higher electronic
and telecommunication technicians

Second edition

D C Green

M Tech, CEng, MIEE

Longman Scientific & Technical,
Longman Group UK Limited,
Longman House, Burnt Mill, Harlow,
Essex CM20 2JE, England
and associated companies throughout the world.

First published in Great Britain by Pitman Books Limited 1981
Reprinted 1983
Reprinted by Longman Scientific & Technical 1986, 1987, 1988
Second edition 1989
Second impression 1991

British Library Cataloguing in Publication Data
Green, D.C. (Derek Charles), 1931−
   Electronics IV.—2nd ed.

   1. Analogue electronic equipment — For
   technicians
   I. Title
   621.381

ISBN 0-582-02183-9

Produced by Longman Group (FE) Ltd
Printed in Hong Kong

CONTENTS

Preface vii

1 Transistors 1
Bipolar transistors 2
The Darlington circuit 12
Field-effect transistors 14
Equivalent circuits 17
Data sheets 31

2 Small-signal Audio-frequency Amplifiers 39
Single-stage transistor amplifiers 39
Multiple stages 51
IC amplifiers 52
Variation of gain with frequency 56
Tuned amplifiers 66
Current sources and current mirrors 68
The differential amplifier 70

3 Negative Feedback 73
Voltage-series feedback 74
Current-series feedback 80
Current-shunt feedback 82
Voltage-shunt feedback 84
Advantages of negative feedback 86
Stability of negative-feedback amplifiers 89

4 Operational Amplifiers 99
Parameters of operational amplifiers 100
Data sheets 109
Inverting amplifier 114
Non-inverting amplifier 119
The voltage follower 121
Compensation of operational amplifiers 122
Selection of an op-amp 124
Differential amplifier 126
Instrumentation amplifier 127
The design of an op-amp circuit 128

5 Audio-frequency Large-signal Amplifiers 131
Single-ended power amplifiers 131
Class B push-pull amplifiers 134
Design of Class B push-pull amplifiers 138
Integrated-circuit power amplifiers 140
Power FET Class B push-pull amplifiers 144

6 Sinusoidal Oscillators 145
The generalized oscillator 146
Resistance−capacitance oscillators 146
Inductance−capacitance oscillators 151
Frequency stability 156
Crystal oscillators 157

7 Non-sinusoidal Waveform Generators 161
The 555 timer 161
Bistable multivibrators 162
Schmidtt trigger 163
Monostable multivibrators 168
Astable multivibrators 174
Ramp and triangle waveform generators 180
Clamping and clipping 189

8 Noise 192
Sources of noise 193
Noise calculations 200
Noise factor (or figure) 203
Noise temperature 208
Measurement of noise figure and noise temperature 209

9 The Phase-locked Loop 211
Components of the phase-locked loop 211
The phase-locked loop 216
Integrated-circuit phase-locked loops 224
Phase-locked loop applications 226

10 Active Filters 229
First-order filters 231
Second-order filters 232
The Butterworth filter 233
The Tchebysheff filter 236
Active filters 239

11 Power Supplies 249
Full-wave rectification 249
Linear voltage regulators 253
Switched-mode power supplies 255
Current limiting and remote sensing 259

Exercises 261

Answers to Numerical Exercises 270

Index 272

Preface

The rapid developments in the use of electronics in all kinds of fields mean that there is a continual demand for people with some knowledge of electronic circuits, devices and techniques. This book provides a comprehensive coverage of the techniques used in modern analogue equipment.

The Business and Technician Education Council has produced a series of units at the second and third levels to introduce the students to the concepts of both analogue (or linear) and digital electronics. At these levels the approach to understanding is mainly descriptive and any analysis is fairly simple and straightforward.

Those students who attain the standard of the Technician Certificate or Diploma and who want to further their studies, then require a more analytical account of circuits, etc. The amount of desirable knowledge is very large and more than can be covered in a standard 60-hour unit, so some selection in the choice of material for a particular course is needed. With this problem in mind the BTEC has produced a large 'unit' covering most aspects of both digital and analogue electronics from which colleges are invited to select material to make up one, or more, 60-hour units.

This book has been written to cover much of the material specified by the *analogue* sections of this large unit. Much of the digital sections are covered by the companion volume entitled *Digital Electronic Technology*.

Chapter 1 introduces the basic principles of transistors. In the following chapters the applications of these devices in a wide variety of circuits are discussed. Chapters 2 to 5 inclusive discuss the ways in which transistors and integrated circuits can be used for the amplification of signals. In chapters 3 and 4 the gain of an amplifier has sometimes been expressed as $20 \log_{10}$ (the voltage gain) dB even when the input and output impedances are (probably) not equal. This is the usual practice when considering Bode diagrams and the stability of feedback amplifiers and does lead to some simplification of problems.

Chapters 6 and 7 are concerned with the methods employed to generate repetitive waveforms. Sinusoidal waveform generators are the subject of Chapter 6, while Chapter 7 deals with the generation of non-sinusoidal waveforms. The important topic of electrical noise is then considered in Chapter 8. Chapters 9, 10 and 11 then deal with phase-locked loops, active filters and power supplies. In a few places, most notably in Chapter 10, some simple use of the Laplace transform has been made. The reader who is not familiar with this technique should merely regard S as being equal to $j\omega$. In many points in the text an indication has been given to the design of a circuit as a basis for the design of more complete systems. Space has not permitted this important aspect of electronics to be further pursued in this book.

The book has been written on the assumption that the reader already possesses a knowledge of Electronics, Electrical Principles and Mathematics of the standard reached by the level III units of the Business and Technician Education Council's scheme for electrical, electronic and telecommunication technicians. Many worked examples are provided in the text to illustrate the principles that have been discussed and a number of exercises are provided at the end of the book. Answers to the numerical exercises are also given.

I wish to express my thanks to Phillips Components Ltd, to National Semiconductor (UK) Ltd, and to RS Components Ltd for their permission to use, and reproduce their copyright material.

DCG

1 Transistors

The transistor is a semiconductor device that performs many varied tasks in electronic circuitry but it is mainly used for amplification and switching. There are two kinds of transistor: the bipolar transistor and the field-effect transistor. Each kind is available both in discrete form and as a component within a monolithic integrated circuit. The desirable parameters of a transistor are set by its intended application. This could be, for example, the amplification of small signals, power amplification and/or control, or the switching of current. The manufacturer's data sheet provides the necessary information on each transistor and this should be consulted when selecting a particular transistor for a specific purpose. Data sheets employ standard symbols to denote various currents, voltages, etc. and these must be clearly understood if the information which is presented is to be fully appreciated.

The system of symbols employed uses the usual symbols I, V, P, etc., to represent current, voltage, power, and so on. Each symbol has a number of suffixes which are written in a definite order; thus

First subscript   denotes the terminal at which the current or voltage is measured,

Second subscript denotes the reference terminal,

Third subscript   indicates whether the third terminal is open, or short-circuited, or connected via a specified resistance R to the reference terminal.

Thus V_{CBO} means the voltage between collector and base with the emitter open-circuited.

Power supply voltages are indicated with two identical suffixes, the first suffix indicates the electrode to which the supply voltage is applied, e.g. V_{CC} is the collector supply voltage.

Transistor capacitances are indicated with the first suffix denoting input or output and the second suffix denoting the transistor connection, i.e. C_{oe} is the output capacitance in the common-emitter connection.

The technical specification of a transistor, obtained from its data sheet, ought not to be the only factor that is considered in the selection of a transistor. Also of importance are such factors as the cost of the transistor, whether or not it is readily available, and whether or not it is *second sourced*. Many devices, integrated circuits as well as transistors, although originally introduced by one manufacturer, are now produced by more than one firm and are said to be second sourced. Clearly, second sourcing is advantageous to the user since it means that the future supplies of that transistor are better assured.

Bipolar Transistors

D.C. Current Gain

The d.c. current gain h_{FE} of a bipolar transistor is the ratio of its d.c. collector current to its d.c. base current, assuming that the collector leakage current is negligibly small. The d.c. current gain is not a constant quantity but varies with both junction temperature and collector current.

Figure 1.1 shows typical variations of d.c. current gain with temperature and with collector current. It is fairly obvious that if the maximum current gain is to be provided by a bipolar transistor it is necessary to bias the device to pass the appropriate value of collector current. Manufacturers' data sheets quote the maximum current gain of a transistor and the value of collector current at which it is obtained. For example, the n-p-n transistor BC 109 has a maximum current gain of 110–450 at a d.c. collector current of 2 mA. The maximum current gain quoted varies widely over the range 110–450. Such a wide variation is characteristic of many types of transistor and arises because of the difficulties associated with any attempts to closely control the manufacturing process. Usually, the d.c. current gain h_{FE} is approximately equal to the small-signal current gain h_{fe}.

When the collector-base junction of a bipolar transistor is reverse biased the collector current is given by

$$I_C = h_{FE}I_B + I_{CEO} \tag{1.1}$$

where I_B is the base current and I_{CEO} is the *collector leakage current* when the transistor is connected with common emitter. I_{CEO} is

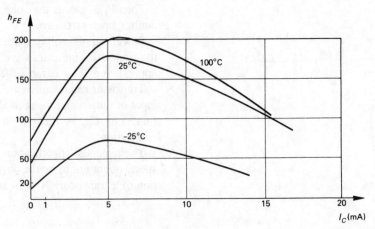

Fig. 1.1 Variation of h_{FE} with collector current and temperature.

related to the *common-base leakage current* I_{CBO} by the expression

$$I_{CEO} = I_{CBO}(1+h_{FE}). \qquad (1.2)$$

For most silicon transistors I_{CBO} is negligibly small — a few nanoamperes — at junction temperatures below about 120 °C. This means that any rise in the collector current due to an increase in I_{CBO} because of a temperature increase is very small, and so any temperature dependence of the collector current is the result of variations in h_{FE}. The effect of an increased value of h_{FE} on the output characteristics of a transistor is to increase the spacing between the curves for each base current. With a germanium transistor I_{CBO} is *not* negligibly small — a few microamperes — and the effect of an increase in the junction temperature is to move the entire family of output characteristic curves upwards.

Transistor data sheets often specify I_{CBO} at room temperatures and sometimes indicate how I_{CBO} varies with change in temperature. When such information is not available it is reasonably accurate to assume that I_{CBO} doubles for every 12 °C rise in temperature for a germanium transistor and for every 8 °C increase for a silicon transistor.

A.C. Current Gain

The a.c. current gain h_{fe} of a bipolar transistor is the ratio $\delta I_C/\delta I_B$ for a constant value of collector-emitter voltage V_{CE}. For collector currents smaller than the value which gives the maximum current gain h_{FE}, $h_{fe} > h_{FE}$ but the reverse is true for larger collector currents. When the collector current is somewhere in the region of the value for maximum h_{FE}, then $h_{fe} \simeq h_{FE}$.

High-frequency Performance

The current gain h_{fe} of a bipolar transistor is not the same at all frequencies but falls at the rate of 6 dB per octave at frequencies above the *cut-off frequency* f_β (see Fig. 1.2). The cut-off frequency is the frequency at which the magnitude $|h_{fe}|$ has fallen by 3 dB from its low-frequency value. The frequency at which $|h_{fe}|$ has fallen to unity is known as the *transition frequency* f_t. At any frequency f less than f_t the magnitude of the current gain can be determined using the relationship

$$f_t = |h_{fe}| f. \qquad (1.3)$$

Transistor data sheets usually specify a minimum value for f_t and very often quote a typical value as well, perhaps with a curve showing how f_t varies with collector current. A typical f_t–collector-current curve is given in Fig. 1.3. Expressions for the cut-off and transition frequencies will be derived later in this chapter (p. 26) after equivalent circuits for transistors have been discussed.

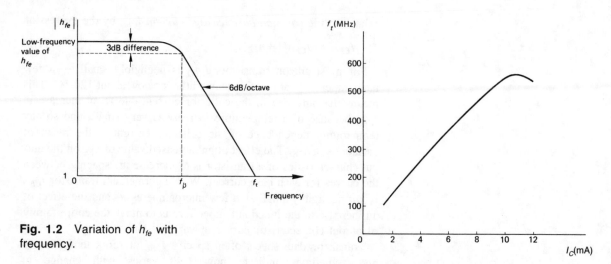

Fig. 1.2 Variation of h_{fe} with frequency.

Fig. 1.3 Variation of f_t with current.

Mutual Conductance

The mutual conductance g_m of a bipolar transistor is the ratio $\delta V_C/\delta I_{BE}$ for a constant value of collector–emitter voltage V_{CE}. Also,

$$g_m = \frac{\delta I_C}{\delta I_B} \times \frac{\delta I_B}{\delta V_{BE}} = \frac{\text{Current gain } h_{fe}}{\text{Input resistance } h_{ie}}. \qquad (1.4)$$

At room temperatures ($T = 300$ K) the value of g_m is approximately given by equation (1.5), i.e.

$$g_m = \frac{\text{quiescent collector current in mA}}{26}$$

$$= 38.5 \text{ mS per mA}. \qquad (1.5)$$

For example, if the quiescent collector current is 2 mA the value of the mutual conductance is approximately 77 mS.

Power Dissipation

An increase in the temperature of the collector–base junction will cause the collector leakage current I_{CBO} to increase. The increased collector current produces an increase in the power dissipated at the junction and this, in turn, further increases the junction temperature and so gives a further increase in I_{CBO}. The process is cumulative and, particularly in the common-emitter connection – since $I_{CEO} \gg I_{CBO}$ – it may damage the device. Excessive heat within a transistor may melt soldered connections, cause the deterioration of insulation materials, and produce chemical and metallurgical changes in the semiconductor material.

The manufacturer of a transistor quotes the maximum permissible power that can be dissipated within the transistor without causing damage. The power dissipated in a transistor is predominantly the power that is dissipated at its collector−base junction and this, in turn, is equal to the d.c. power taken from the collector supply voltage minus the total output power (d.c. power plus a.c. power). For transistors handling small signals, the power dissipated at the collector is small and there is generally little problem. When the power dissipated by the transistor is large enough to cause the junction temperature to rise to a dangerous level, it is necessary to increase the rate at which heat is removed from the device. Power transistors are constructed with their collector terminal connected to their metallic case. To increase the area from which the heat is removed, the case of the transistor can be bolted on to a sheet of metal which acts as a heat sink. Heat will move from the transistor to the heat sink by conduction and then be removed from the heat sink by convection and radiation.

The heat dissipated at the collector−base junction flows through the thermal resistance θ between the junction and the external environment. Thermal resistance is analogous to electrical resistance and represents the opposition of a material to the flow of heat energy. Under steady-state conditions the thermal resistance of a thermal conductor is the ratio of the temperature drop across the conductor to the heat transfer rate (in joules/second or watts) through the conductor. The unit of thermal resistance is 'degrees centigrade per watt'. When a continuous power is dissipated at the collector−base junction, the heat generated will cause the junction temperature to rise until the difference between the junction and ambient temperatures is equal to the product of the thermal resistance and the power dissipated at the junction, i.e.

$$T_J - T_A = \theta_{JA} P_c \tag{1.6}$$

T_J is the safe junction temperature. This figure is given by the manufacturer and is usually about 85 °C for a germanium transistor and about 150 °C for a silicon transistor.

T_A is the ambient temperature and may be higher than the room temperature since the transistor may be installed within an enclosed space that contains one or more other heat-producing devices.

The thermal resistance consists of three parts.

1   The thermal resistance θ_{JC} between the collector−base junction and the case of the transistor.
2   The thermal resistance θ_{CS} between the transistor case and the heat sink.
3   The thermal resistance θ_{SA} between the heat sink and the environment at ambient temperature.

The conduction of heat energy is analogous to the conduction of electricity and, for calculations on heat performance, electrical analogies are used. Thus,

Heat flow $\equiv$ current flow
Temperature difference $\equiv$ potential difference
Thermal resistance $\equiv$ electrical resistance.

Using these analogies, the equivalent circuit for the thermal performance of a transistor can be drawn (see Fig. 1.4).

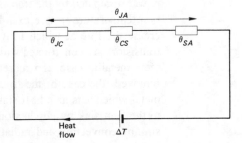

Fig. 1.4 Thermal equivalent circuit of a transistor and its heat sink.

The thermal resistance θ_{JC} between the collector−base junction and the transistor case is normally quoted by the manufacturer. The required total thermal resistance for a given power dissipation can be calculated and then a suitable heat sink can be selected with the aid of the heat-sink data available from makers.

Example 1.1

A transistor has a maximum collector dissipation of 6 W. The maximum junction temperature is 80 °C, $\theta_{JC} = 0.9$ °C/W and $\theta_{CS} = 0.25$ °C/W. Calculate the necessary thermal resistance of the heat sink if the ambient temperature is 20 °C.

Solution

$$\theta_{JA} = \frac{T_J - T_A}{P_c} = \frac{80 - 20}{6} = \frac{60}{6} = 10 \text{ °C/W}$$

Therefore,

$$\theta_{SA} = 10 - 0.9 - 0.25 = 8.85 \text{ °C/W.} \quad (Ans.)$$

The maximum power dissipation figure quoted by a manufacturer usually assumes a maximum ambient temperature of 25 °C. If the ambient temperature is likely to be higher than 25 °C, the maximum collector dissipation must be reduced. Sometimes the necessary derating is given as a figure, e.g. 'derate 4 mW/°C for ambient temperatures above 25 °C' but often a *derating curve* is given. A typical derating curve is shown in Fig. 1.5; the slope of the curve is equal to the derating factor, i.e. $-240/(85-25)$ or -4 mW/°C.

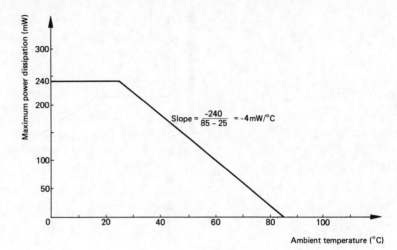

Fig. 1.5 Transistor dissipation derating curve.

The rise in junction temperature is equal to the power dissipated in the transistor divided by the derating factor. This means that the thermal resistance is equal to the reciprocal of the derating factor, and hence for the transistor referred to in Fig. 1.5, $\theta_{JA} = 250 \, °C/W$.

Example 1.2

The following data is provided for a bipolar transistor:

Maximum power dissipation 12 W at 25 °C
Maximum junction temperature 120 °C
Thermal resistance 8 °C/W

Calculate the maximum power dissipation for an ambient temperature of 80 °C.

Solution
Method 1 $P_c = \dfrac{T_J - T_A}{\theta_{JA}} = \dfrac{120 - 80}{8} = 5 \, W$ (*Ans.*)

Method 2   The derating curve for the transistor falls linearly from 12 W at $T_A = 25 \, °C$ to 0 W at $T_A = 120 \, °C$ (see Fig. 1.6). From this graph, when $T_A = 80 \, °C$, $P_d = 5 \, W$ (*Ans.*)

The maximum collector dissipation of a transistor can be marked on the output characteristics by drawing the locus of the power dissipation as the collector current varies.

Thermal Time Constant

The transistor, its case, and its heat sink must all store, or release, heat energy in order to be able to change in temperature. The effect

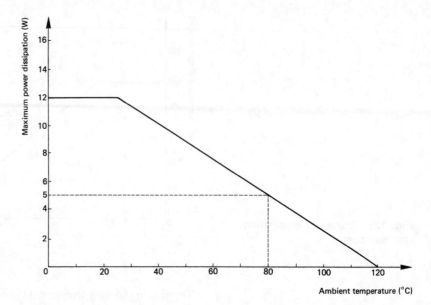

Fig. 1.6

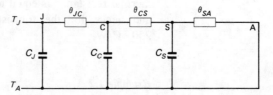

Fig. 1.7 Thermal equivalent circuit with thermal 'capacitance'.

is analogous to electrical capacitance which must store, or release, electric charge in order to change in voltage. The *thermal time constant* is the product of thermal resistance and thermal 'capacitance'. The electrical equivalent circuit is shown in Fig. 1.7 where C_J, C_C and C_S are, respectively, the thermal capacitances of the junction, case and heat sink.

Example 1.3

The case temperature of a transistor increases by 28.5 °C after 1 minute, and by 42.75 °C after 2 minutes of operation. The maximum junction temperature is 110 °C. The thermal resistances are

Junction-case 0.8 °C/W
Case-heat sink 0.5 °C/W
Case-air 3.0 °C/W

If the ambient temperature is 25 °C calculate (*a*) the maximum transistor dissipation, (*b*) the necessary thermal resistance of the heat sink.

Solution

Let T_m be the final increase in the case temperature. Then

$$28.5 = T_m(1-e^{-1/\tau})$$
$$42.75 = T_m(1-e^{-2/\tau})$$

where τ is the thermal time constant in minutes.

Multiplying the first equal equation by $(1+e^{-1/\tau})$ gives

$$28.5(1+e^{-1/\tau}) = T_m(1-e^{-2/\tau}) = 42.75.$$

Therefore

$$e^{-1/\tau} = \frac{42.75}{28.5} - 1 = \tfrac{1}{2}$$

and hence

$$28.5 = T_m(1-\tfrac{1}{2}) \quad \text{or} \quad T_m = 57 \text{ °C}.$$

Therefore

Final case temperature $= 57$ °C $+ 25$ °C $= 82$ °C.

Maximum collector dissipation

$$P_c = \frac{T_J - T_C}{\theta_{JC}} = \frac{110-82}{0.8} = 35 \text{ W}. \hspace{2cm} (Ans.)$$

Also $P_c = \dfrac{T_J - T_A}{\theta_T}$ so

$$\theta_T = \frac{T_J - T_A}{P_c} = \frac{110-25}{35} = 2.43 \text{ °C/W}.$$

θ_T is the *total* thermal resistance between the transistor junction and the surrounding air. The thermal resistance 'circuit' is shown in Fig. 1.8, from which

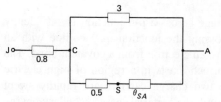

Fig. 1.8

$$2.43 = 0.8 + \frac{3(0.5+\theta_{SA})}{3.5+\theta_{SA}}$$

$$8.5+2.43\theta_{SA} = 2.8+0.8\theta_{SA}+1.5+3\theta_{SA} \quad \theta_{SA} = 3 \text{ °C/W}.(Ans.)$$

Maximum Collector Current

The maximum collector current that should be allowed to flow in a bipolar transistor is limited by the power handling capability of the internal connections of the various regions. If the maximum safe current is exceeded, the connections may be damaged, semiconductor materials may change and the current gain may be dramatically reduced.

Data sheets usually quote both the maximum quiescent collector current I_C and the maximum collector current, I_{CM}, with a signal. The base current needed to give this I_{CM} value is not generally quoted. In a design, the collector current is usually limited to some 60–70% of the quoted maximum.

Maximum Collector Voltage

If the voltage applied across a p-n junction exceeds the *breakdown* value, a large current will flow which may cause damage to the transistor. The maximum safe voltages are generally quoted as between either (*a*) the collector and base terminals with the emitter open-circuit or (*b*) the collector and emitter terminals with the base open-circuit. These two voltages are indicated by the labels V_{CBO} and V_{CEO}, respectively. Typically, $V_{CBO} = 40$ V, $V_{CEO} = 60$ V.

There are two possible causes of the voltage breakdown of a bipolar transistor.

Punch-through

Increasing the collector−emitter voltage V_{CE} of a transistor will increase the reverse bias of the collector−base junction and the forward bias of the emitter−base junction, and in so doing will reduce the effective width of the base region. Eventually, the base width will approach zero and the current flowing will increase to a large value since the transistor has effectively become a low resistance connected across the power supply. The heat developed by this large current will damage the transistor.

Avalanche Multiplication

When the reverse-bias voltage across a p-n junction is sufficiently large, a charge carrier crossing the junction may collide with an atom and in so doing remove an electron from a covalent bond. The hole−electron pair thus formed exists in a region of high electric field strength, and so the two charge carriers are rapidly swept away. In their passage through the material these secondary charge carriers may, in turn, also collide with atoms to produce further hole−electron pairs. This process is cumulative and will lead to a rapid increase in the collector current if the voltage is high enough.

For a given transistor, voltage breakdown might be due to either punch-through or avalanche multiplication, depending on the structure of the device. The voltage V_{EBO} applied between base and emitter must also be kept below a quoted maximum figure.

The Safe Operation of Power Transistors

The main ratings of a power transistor are based upon the maximum permissible values of collector current, collector voltage, power dissipation and junction temperature. The *safe operating area* (SOAR) of a bipolar transistor is a plot of the log of collector current to a base of the log of the collector−emitter voltage on which four limits are marked. Figure 1.9 shows a typical SOAR diagram. Three

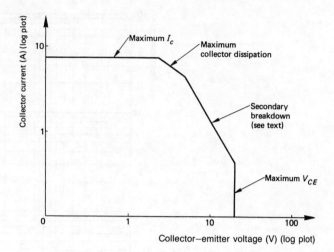

Fig. 1.9 SOAR diagram for a transistor.

of the boundaries to the diagram have already been discussed; the fourth, namely *secondary breakdown*, occurs if the collector current is allowed still further to increase after avalanche breakdown has taken place. Secondary breakdown results in an excessive temperature rise somewhere within the transistor and will usually destroy the device. When designing a circuit, each of the four SOAR boundaries should be regarded as a limit which must not be exceeded.

Because of secondary breakdown the rated maximum power dissipation applies only for voltages up to about 10% of the rated value of V_{CEO}. At the rated V_{CEO} value the possible power dissipation is only about 5% of the rated maximum.

Example 1.4

A transistor has a maximum collector current of 10 A and maximum collector−emitter voltage V_{CEO} of 80 V. The maximum power dissipated at an ambient temperature of 25 °C is 40 W. Draw the SOAR diagram. Assume that the maximum collector current at $V_{CE} = 80$ V before the onset of secondary breakdown is 0.3 A.

Solution

Since the maximum collector dissipation is 40 W the relationship between I and V is

I_C	(A)	1	2	4	8	10
V_{CE}	(V)	40	20	10	5	4

Plotting these figures on log/log axes gives a straight line with a slope of -1 (Fig. 1.10).

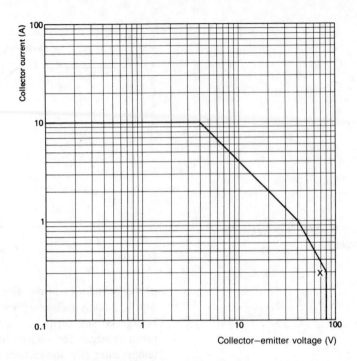

Fig. 1.10

The maximum collector current at $V_{CE} = 80$ V is 0.3 A; this point is marked as X in Fig. 1.10. A straight line must now be drawn, with a slope greater than unity, to connect point X to the maximum dissipation line. The slope of this line must be such that the reduction in collector−emitter voltage is large enough to prevent secondary breakdown. A suitable line is shown in the figure.

Transistor Capacitances

Each of the p-n junctions in a bipolar transistor has a self-capacitance, the value of which is dependent upon both the magnitude and polarity of the junction voltage. This capacitance represents the charge stored when a voltage is applied.

The transistor capacitances are quoted in various ways by different manufacturers:

(*a*) collector−base capacitance C_{Tc} and emitter−base capacitance C_{Te}, and
(*b*) common-base input and output capacitances C_{ib} and C_{ob}.

When the transistor is operated in the common-emitter connection C_{ob} is the collector−base, and C_{ib} is the emitter−base capacitance.

The Darlington Circuit

The Darlington connection, illustrated by Fig. 1.11, has two transistors with their base and collector terminals commoned to produce an element which can often be used to replace a single transistor.

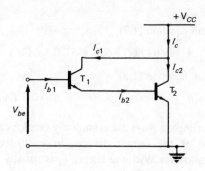

Fig. 1.11 Darlington connection.

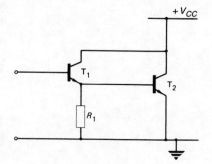

Fig. 1.12 Use of an emitter resistor in a Darlington pair.

From Fig. 1.11,

$$I_{c1} = h_{fe1}I_{b1} \quad \text{and} \quad I_{e1} = I_{b2} = I_{b1}(1+h_{fe1})$$

Also

$$I_{c2} = h_{fe2}I_{b2} = h_{fe2}I_{b1}(1+h_{fe1}).$$

The *total* collector current I_c is the sum of the individual collector currents, i.e.

$$I_c = h_{fe1}I_{b1} + h_{fe2}I_{b1}(1+h_{fe1})$$

and the short-circuit current gain is

$$A_i = I_c/I_{b1} = h_{fe1} + h_{fe2}(1+h_{fe1}). \qquad (1.7)$$

The current gain of a Darlington-connected pair of transistors is much greater than that provided by either transistor alone.

Also, from Fig. 1.11,

$$V_{be} = I_{b1}R_{in1} + (1+h_{fe1})I_{b1}R_{in2}$$

where R_{in1} and R_{in2} are, respectively, the input resistances of T_1 and T_2. Therefore,

$$R_{int} = V_{be}/I_{b1} = R_{in1} + (1+h_{fe1})R_{in2}. \qquad (1.8)$$

Equation (1.8) shows that the input resistance of a Darlington pair can be very high. Thus, the two main characteristics of a Darlington pair are a very high current gain and very high input impedance.

The emitter current of T_1 must always be larger than the collector leakage current of T_2 and, to ensure this, it is often necessary to connect an emitter resistor in the position shown in Fig. 1.12.

Example 1.5

Calculate the current gain and input resistance of a Darlington pair if identical transistors are used having $h_{fe}=60$ and $h_{ie}=1000 \ \Omega$. Assume h_{re} and h_{oe} can be neglected.

Solution

From equation (1.7)

$$A_i = 60 + 60(1 + 60) = 3720. \quad (Ans.)$$

From equation (1.8)

$$R_{int} = 1000 + 61 \times 1000 = 62 \text{ k}\Omega. \quad (Ans.)$$

Darlington pairs are commonly employed within integrated circuits, and as the output devices in audio-frequency push−pull power amplifiers. With the latter, T_1 is usually a low-power high-current-gain type and T_2 is the power transistor proper which passes a large current. Difficulties can arise with the circuit if the collector current of the second transistor is low and its gain is fairly large. Suppose, for example, that T_2 has $h_{fe} = 150$ and $I_c = 2$ mA. Then

$$I_{b2} \simeq I_{e1} \simeq I_{c1} = 2 \times 10^{-3}/150 = 13.33 \ \mu\text{A}$$

and for such a low value of collector current the gain of T_1 will be small.

This would mean that much of the advantage of the Darlington connection would be lost. Darlington pairs are available as discrete packages. Power Darlington transistors have a good frequency response, are easy to use, and are generally cheaper than power MOSFETs. One device, the TIP 646 has the following data: maximum power disssipation 100 W, maximum current 10 A, maximum V_{CEO} 80 V, and current gain $h_{fe} = 1000$ at $I_C = 5$ A.

Field-effect Transistors

The field-effect transistor (FET) is a semiconductor device which can perform most of the functions of the bipolar transistor but which operates in a fundamentally different way. Three kinds of FET are available: the junction FET or JFET, the metal oxide semiconductor FET or MOSFET and the power MOSFET. The small-signal MOSFET is, in turn, divisible into two types: the *depletion* type and the *enhancement* type. The important characteristics of a FET are its

Mutual conductance $g_m = \delta I_D/\delta V_{GS}$ with V_{DS} constant
Input resistance R_{in}
Drain−source resistance $r_{ds} = \delta V_{DS}/\delta I_D$.

Typically, $g_m = 4$ mS, $r_{ds} = 100$ kΩ and $R_{in} = 100$ MΩ plus for a JFET and 10^{10} Ω for a MOSFET.

The drain current of a FET depends upon both the gate−source voltage V_{GS} and the drain−source voltage V_{DS}. For a junction FET the drain current which flows when $V_{GS} = 0$ is the maximum drain current and it is labelled as I_{DSS}. For a depletion-mode MOSFET the current I_{DSS} which flows for $V_{GS} = 0$ is not the maximum drain current.

For both a junction FET and a depletion-mode MOSFET, two important parameters are

(a) the gate pinch-off voltage V_P

(b) the drain current I_{DSS} which flows when $V_{GS} = 0$.

When a JFET is operated in its pinch-off region, i.e. as an amplifying device,

$$I_D = I_{DSS}(1-|V_{GS}/V_P|)^2. \tag{1.9}$$

The mutual conductance g_m is given by $\delta I_d/\delta V_{GS}$, hence differentiating equation (1.9)

$$g_m = dI_D/dV_{GS} = \frac{-2I_{DSS}}{V_P}(1-|V_{GS}/V_P|)^2. \tag{1.10}$$

When V_{GS} is zero

$$g_m = g_{mo} -2I_{DSS}/V_P. \tag{1.11}$$

Typically, $I_{DSS} = 1$ mA and $V_P = -2$ V whence $g_{mo} = 1$ mS.

The mutual conductance varies linearly with the gate−source voltage, but it is always very much smaller than the mutual conductance of a bipolar transistor. This means that the FET will generally give a much lower voltage gain than is available from a bipolar transistor.

For an enhancement-type MOSFET the drain current I_D is related to the gate−source voltage V_{GS} by

$$I_D = k(V_{GS}-V_{TH})^2. \tag{1.12}$$

Differentiating,

$$g_m = 2k/(V_{GS}-V_{TH}) \tag{1.13}$$

where k is a constant, and V_{TH} is the gate−source voltage at which the induced channel just becomes evident. V_{TH} is often known as the *threshold voltage*. Typical values for V_{TH} are 1 V to 5 V; when the gate voltage is less than V_{TH} there is no channel and $I_D = 0$. k is usually about 0.3.

Effects of Temperature

The velocity with which majority charge carriers travel through the channel is dependent upon both the drain−source voltage and the temperature of the FET. An increase in the temperature reduces the velocity of the channel charge carriers and this appears in the form of a reduction in the drain current which flows for given gate−source and drain−source voltages.

A further factor that may also affect the variation of drain current with change in temperature is the barrier potential across the gate−channel p-n junction. An increase in temperature will cause the barrier potential to fall and this, in turn, will reduce the width of the depletion layer for a given gate−source voltage. The channel resistance will fall and so the drain current will increase.

The two effects tend to vary the drain current in opposite directions and as a result the overall variation can be quite small. Indeed, if the gate–source voltage is made equal to the pinch-off voltage *plus* 0.63 V then zero temperature coefficient is obtained. In general, the overall result is that the drain current decreases with increased temperature. This is the inverse of the variation in collector current with temperature experienced by a bipolar transistor.

Power Dissipation

The total power dissipated within a FET is the product of the drain current I_D and the drain–source voltage V_{DS}. The manufacturer of a device specifies a maximum safe value for the power dissipation which should not be exceeded.

For a small-signal FET the allowable power dissipation is in the region of 200–300 mW.

Maximum Drain–Source Voltage

If the drain–source voltage of a FET is increased to too large a value, a large increase in the drain current will suddenly take place. There are two contributory reasons for this: (*a*) avalanche breakdown of the reverse-biased drain-substitute p-n junction, and (*b*) punch-through across the channel. The avalanche breakdown effect is similar to that already described for the bipolar transistor.

Punch-through occurs when the reverse-bias voltage at the drain-substrate p-n junction is large enough for the depletion layer thus formed to widen to such an extent that it reaches the source. The source-substrate p-n junction will then break down so that the drain and source terminals of the FET are effectively short-circuited together. The manufacturer of a device quotes the maximum drain–source voltage that can be applied without danger of breakdown taking place.

For most small-signal FETs, the maximum drain–source voltage is 25–30 V.

Reverse Gate–Source Voltage

There is always a maximum value to the reverse gate–source voltage which can be applied without damaging the device. For a JFET, the gate-channel junction must be held in the reverse-biased condition or else a gate current will flow and the input resistance will fall to a low value, probably damaging the device. On the other hand the reverse bias voltage must not be too large or the junction will break down. The gate-channel junction of either type of MOSFET can be either forward biased or reverse biased but again damage can be caused if too large a gate—source voltage is applied. Typically, $V_{GS(max)}$ is 30 V.

Maximum permissible values may also be given by a manufacturer for the gate—drain voltage V_{GD}, and the voltages between each terminal and the substrate, e.g. V_{DB}, V_{SB} and V_{GB}.

FET Capacitances

A FET has capacitances between its gate and source terminals, and between its gate and drain terminals, labelled respectively as C_{GS} and C_{GD}. Data sheets usually quote values for capacitances C_{ISS} and C_{RSS} where, C_{ISS} is the common-source short-circuit input capacitance and is numerically equal to the sum of C_{GS} and C_{GD} and C_{RSS} is the common-source reverse-transfer capacitance which is equal to C_{GD}. Sometimes the output capacitance C_{OSS} is also quoted. Typical values are $C_{ISS} = 5$ pF, $C_{RSS} = 1$ pF, and $C_{OSS} = 3$ pF.

Power MOSFETs

A range of power MOSFETs, known variously as VMOS, DMOS, TMOS or HEXFET, provide an alternative to the bipolar transistor for many power applications. When the bipolar transistor is conducting it dissipates less power than does a power MOSFET but the reverse is true when a switching application is considered. This means that for power switching applications the power MOSFET appears to give a superior performance at frequencies greater than about 40 kHz. The power MOSFET does not suffer from secondary breakdown problems like the bipolar transistor and such a slope is not shown on its SOAR curve. Consequently, the power MOSFET is the more rugged device and is better able to withstand overloads. Because it does not require an input current the drive circuitry for a power MOSFET tends to be simpler than that for a power bipolar transistor. The power bipolar transistor remains superior on cost grounds for applications using voltages below about 500 V and at the lower frequencies. The main fields of application for the power MOSFET are switched-mode power supplies, audio power amplifiers, motor speed control and induction heating.

Equivalent Circuits

An equivalent circuit is one that behaves electrically in exactly the same way as the device whose a.c. performance it represents. The equivalent circuit is only valid over a limited frequency range and provided the device is, or can reasonably be supposed to be, linear in its operation. The values of the components of the equivalent circuit are only valid at the specified collector or drain current and frequency, and provided the signal amplitude is small.

Bipolar Transistor Equivalent Circuits

Several possible equivalent circuits for the bipolar transistor can be drawn but only a few are used. For audio-frequency work, the *h-parameter* equivalent circuit is generally employed although the *hybrid-π* circuit is also sometimes used. At higher frequencies the hybrid-π and the *y*-parameter equivalent circuits are available, while for very high frequency work *s*-parameters provide the greatest accuracy.

h-parameters

Fig. 1.13 Representation of a transistor as a 'black box'.

The performance of a common-emitter transistor may be specified in terms of a general four-terminal network that is itself defined in terms of the currents and voltages existing at its input and output terminals (Fig. 1.13). If the input current I_b and the output voltage V_{ce} are taken as the independent variables, then the output current I_c and the input voltage V_{be} are given by these equations

$$V_{be} = h_{ie}I_b + h_{re}V_{ce} \tag{1.14}$$

$$I_c = h_{fe}I_b + h_{oe}V_{ce}. \tag{1.15}$$

Equation (1.14) states that the input voltage V_{be} is equal to a parameter h_{ie} times the input current I_b plus another parameter h_{re} times the output voltage V_{ce}. The right-hand side of the equation must have the dimensions of voltage, and hence h_{ie} must be an impedance and h_{re} must be dimensionless. Similarly, the right-hand side of equation (1.15) must have the dimensions of a current; therefore h_{fe} is dimensionless and h_{oe} is an admittance.

The a.c. equivalent circuit that is suggested by equations (1.14) and (1.15) is shown in Fig. 1.14; the circuit is equally valid for the common-base and common-collector configurations, the configuration concerned being indicated by a different second suffix, *b* or *c*. The *h*-parameters of a transistor are defined in Table 1.1.

Relationships between h-parameters

Manufacturers do not normally publish transistor data for all three configurations and it is sometimes necessary to know the relationships given in Table 1.2.

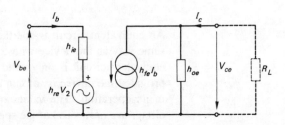

Fig. 1.14 *h*-parameter equivalent circuit of a transistor.

Table 1.1 h-parameters of a transistor

Common base	h-parameter Common emitter	Common collector	Definition
$h_{ib} = \dfrac{V_{eb}}{I_e}$	$h_{ie} = \dfrac{V_{be}}{I_b}$	$h_{ic} = \dfrac{V_{bc}}{I_b}$	Input impedance with output terminals short-circuited (ohms)
$h_{rb} = \dfrac{V_{eb}}{V_{cb}}$	$h_{re} = \dfrac{V_{be}}{V_{ce}}$	$h_{rc} = \dfrac{V_{bc}}{V_{ec}}$	Reverse voltage ratio with input terminals open-circuited (dimensionless)
$h_{fb} = \dfrac{I_c}{I_e}$	$h_{fe} = \dfrac{I_c}{I_b}$	$h_{fc} = \dfrac{I_e}{I_b}$	Forward current gain with output terminals short-circuited (dimensionless)
$h_{ob} = \dfrac{I_c}{V_{cb}}$	$h_{oe} = \dfrac{I_c}{V_{ce}}$	$h_{oc} = \dfrac{I_e}{V_{ec}}$	Output admittance with input terminals open-circuited (siemens)

Table 1.2 h-parameter relationships

Common base	Common emitter	Common collector
h_{ib}	$h_{ie} = \dfrac{h_{ib}}{1 + h_{fb}}$	$h_{ic} = \dfrac{h_{ib}}{1 + h_{fb}}$
h_{rb}	$h_{re} = \dfrac{h_{ib}h_{ob}}{1 + h_{fb}} - h_{rb}$	$h_{rc} \approx 1$
h_{fb}	$h_{fe} = \dfrac{-h_{fb}}{1 + h_{fb}}$	$h_{fc} = \dfrac{-1}{1 + h_{fb}}$
h_{ob}	$h_{oe} = \dfrac{h_{ob}}{1 + h_{fb}}$	$h_{oc} = \dfrac{h_{ob}}{1 + h_{fb}}$

The h-parameters of a transistor are not constant but vary with change both in temperature and in collector current. Figure 1.15 shows how the four h-parameters may vary for a typical transistor. Typical values for the four h-parameters are

$$h_{ie} = 1000 \ \Omega \quad h_{re} = 3 \times 10^{-4} \quad h_{oe} = 300 \times 10^{-6} \text{S} \quad h_{fe} = 250$$

Current Gain
An increase in the collector current will produce a larger voltage drop across the load resistance R_L and so the collector–emitter voltage will fall. Thus $V_{ce} = -I_c R_L$.

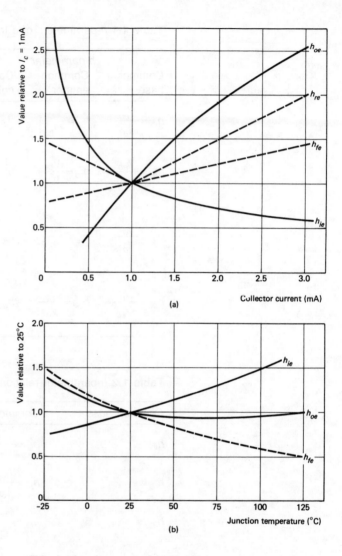

Fig. 1.15 Variation of transistor
h-parameters with
(*a*) collector current,
(*b*) temperature.

Substituting into equation (1.15),

$$I_c = h_{fe}I_b - h_{oe}I_cR_L$$

$$I_c(1 + h_{oe}R_L) = h_{fe}I_b$$

$$I_c = \frac{h_{fe}I_b}{1 + h_{oe}R_L}$$

so that the current gain A_i is

$$A_i = I_c/I_b = \frac{h_{fe}}{1 + h_{oe}R_L}. \tag{1.16}$$

Input Resistance
The input resistance of a bipolar transistor is the ratio of the input
voltage to the input current, i.e.

$$R_{in} = V_{be}/I_b \ \Omega.$$

Equation (1.14) can be written as

$$V_{be} = h_{ie}I_b - h_{re}I_cR_L$$

$$= h_{ie}I_b - \frac{h_{re}h_{fe}R_LI_b}{1+h_{oe}R_L}$$

so that the input resistance is

$$R_{in} = V_{be}/I_b = h_{ie} - \frac{h_{re}h_{fe}R_L}{1+h_{oe}R_L}. \tag{1.17}$$

Voltage Gain

An expression for the voltage gain can be obtained by substituting both (1.16) and (1.17) into $A_v = A_iR_L/R_{in}$ to give

$$A_v = \frac{h_{fe}R_L}{1+h_{oe}R_L} \times \frac{1+h_{oe}R_L}{h_{ie}+h_{ie}h_{oe}R_L-h_{re}h_{fe}R_L}$$

$$= \frac{h_{fe}R_L}{h_{ie}+h_{ie}h_{oe}R_L-h_{re}h_{fe}R_L}. \tag{1.18}$$

Very often h_{re} is negligibly small, and then equations (1.17) and (1.18) reduce to

$$R_{in} = h_{ie} \tag{1.19}$$

$$A_v = \frac{h_{fe}R_L}{h_{ie}(1+h_{oe}R_L)} \tag{1.20}$$

If the product $h_{oe}R_L$ is small compared to unity, h_{oe} may also be neglected with the introduction of an error which will normally be no greater than that due to resistor and transistor tolerances. If h_{oe} is neglected $A_v = g_mR_L$.

Example 1.6

A transistor connected with common emitter has $h_{ie} = 4500 \ \Omega$, $h_{oe} = 50 \times 10^{-6} \text{S}$ and $h_{fe} = 300$. Calculate the voltage and power gains obtained with a collector load resistor of 2000 Ω. Find also the percentage error if h_{oe} is neglected.

Solution

$$A_i = 300/(1+50 \times 10^{-6} \times 2000) = 273$$

$$R_{in} = 4500 \ \Omega$$

Therefore,

$$A_v = 273 \times 2 \times 10^3/4500 = 121. \quad (Ans.)$$

If h_{oe} is neglected, $A_i = h_{fe} = 300$ and

$$A_v = 300 \times 2 \times 10^3/4500 = 133.$$

The percentage error in the calculated voltage gain is

$$\frac{133-121}{121} \times 100 \simeq +10\%. \quad (Ans.)$$

The error may seem rather large but it is of the same order as the probable errors caused by resistor tolerances ($\pm 5\%$ or $\pm 10\%$), and also the transistor parameters have wide tolerances.

Output Resistance

The *output resistance* R_{out} is the ratio V_{ce}/I_c with zero input voltage.

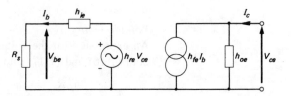

Fig. 1.16 Calculation of R_{out}.

From Fig. 1.16

$$V_{be} = -I_b R_s.$$

Therefore, equation (1.14) becomes

$$-I_b R_s = h_{ie} I_b + h_{re} V_{ce} \quad \text{and} \quad I_b = \frac{-h_{re} V_{ce}}{h_{ie}+R_s}.$$

Then, equation (1.15) becomes

$$I_c = \frac{-h_{fe} h_{re} V_{ce}}{h_{ie}+R_s} + h_{oe} V_{ce}$$

$$R_{out} = V_{ce}/I_c = \frac{h_{ie}+R_s}{h_{oe} h_{ie}+h_{oe} R_s - h_{fe} h_{re}} \qquad (1.21)$$

Note that the output resistance of a bipolar transistor depends upon the value of the source resistance. However, if h_{re} is negligibly small,

$$I_c = h_{oe} V_{ce} \quad \text{and} \quad R_{out} = \frac{1}{h_{oe}}. \qquad (1.22)$$

Hybrid-π Equivalent Circuit

The *h*-parameter equivalent circuit is normally only used at audio frequencies where the self-capacitances of the transistor have negligible effect on the performance of the device. At higher frequencies the effects of the transistor capacitances cannot be ignored and so these capacitances must appear in any equivalent circuit of the transistor. One circuit, which includes the transistor capacitances

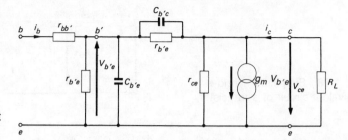

Fig. 1.17 Hybrid-π equivalent circuit of a transistor.

in a way which takes account of the manner in which the current gain of the transistor varies with frequency, is known as the hybrid-π circuit and is shown in Fig. 1.17. The point labelled b' is supposed to be in the middle of the base region. The meaning of each of the components shown is as follows:

g_m = mutual conductance I_c/V_{be} of the transistor.

$r_{bb'}$ = ohmic resistance of the base region: typically $r_{bb'} = 200 \ \Omega$.

$r_{b'e}$ = resistance of the forward biased base–emitter junction

$$r_{b'e} = V_{b'e}/I_b = V_{b'e}/I_c \times I_c/I_b = h_{fe}/g_m$$

$r_{b'e}$ is typically 1000 Ω.

$C_{b'e}$ = capacitance of the forward biased base–emitter junction, and is about 20 pF.

$C_{b'c}$ = capacitance of the reverse biased collector–emitter junction, typically 10 pF.

$r_{b'c}$ = resistance of the reverse biased collector–base junction, typically 1 MΩ. Since $r_{b'c}$ has so high a value it can usually be omitted from the equivalent circuit without introducing undue error.

r_{ce} = output resistance of the transistor when the base–emitter terminals are short-circuited and is typically about 20 kΩ.

It is always possible to convert from one equivalent circuit to another and Table 1.3 gives the relationships for an h-parameter to hybrid-π conversion.

Voltage Gain

At low frequencies where the capacitances $C_{b'e}$ and $C_{b'c}$ can be neglected and removed from the equivalent circuit (Fig. 1.18), a voltage V_{be} applied between the base and emitter produces

$$V_{b'e} = V_{be}r_{b'e}/(r_{bb'} + r_{b'e}).$$

Table 1.3

Equivalent circuit	Parameter			
	h_{ie}	h_{re}	h_{fe}	h_{oe}
h				
hybrid-π	$r_{bb'} + r_{b'e}$	$r_{b'e}/r_{b'c}$	$g_m r_{b'e}$	$1/r_{ce}$

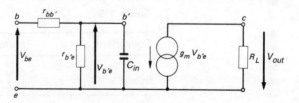

Fig. 1.18 Simplified hybrid-π equivalent circuit of a transistor.

The output voltage V_{ce} is

$$V_{ce} = \frac{g_m V_{be} r_{b'e}}{r_{bb'} + r_{b'e}} \times \frac{R_L r_{ce}}{R_L + r_{ce}}.$$

Usually, $R_L \ll r_{ce}$ and then the voltage gain A_v is

$$A_v = V_{ce}/V_{be} = \frac{g_m r_{b'e} R_L}{r_{bb'} + r_{b'e}} \simeq g_m R_L. \tag{1.23}$$

At high frequencies the reactances of $C_{b'e}$ and $C_{b'c}$ are small enough to affect the current and voltage gains of the transistor. The effect of the capacitance $C_{b'c}$ linking the input and output circuits of the transistor is to considerably increase the input capacitance of the device.

The current i flowing through $C_{b'c}$ is

$$i = j\omega C_{b'c}(V_{b'e} - V_{ce})$$
$$= j\omega C_{b'c}(V_{b'e} - [-g_m R_L V_{b'e}])$$
$$= j\omega C_{b'c} V_{b'e}(1 + g_m R_L).$$

Thus, the *admittance* seen looking into the left-hand side of $C_{b'c}$ is

$$Y_{in} = i/V_{b'e} = j\omega C_{b'c}(1 + g_m R_L).$$

The effective input capacitance C_{in} of the transistor is the sum of $C_{b'e}$ and the capacitance represented by Y_{in}, i.e.

$$C_{in} = C_{b'e} + C_{b'c}(1 + g_m R_L). \tag{1.24}$$

The increase in the input capacitance brought about in this way is known as the *Miller Effect*. The simplified high-frequency equivalent circuit of the transistor is shown by Fig. 1.19. From this circuit the low-frequency voltage gain A_v is unchanged at $g_m r_{b'e} R_L/(r_{bb'} + r_{b'e})$.

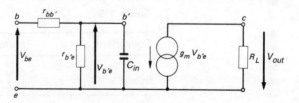

Fig. 1.19 High-frequency hybrid-π equivalent circuit.

At high frequencies, however, the voltage $V_{b'e}$ is given by

$$V_{b'e} = \frac{V_{be}r_{b'e}/(1+j\omega C_{in}r_{b'e})}{r_{bb'}+r_{b'e}/(1+j\omega C_{in}r_{b'e})}$$

$$= \frac{V_{be}r_{b'e}}{r_{bb'}+r_{b'e}+j\omega C_{in}r_{b'e}r_{bb'}}$$

Since

$$V_{ce} = g_m V_{b'e} R_L$$

$$A_{v(hf)} = V_{ce}/V_{be} = \frac{g_m r_{b'e} R_L}{r_{bb'}+r_{b'e}+j\omega C_{in}r_{b'e}r_{bb'}}. \tag{1.25}$$

Equation (1.25) can be rewritten as

$$A_{v(hf)} = \frac{g_m r_{b'e} R_L}{r_{bb'}+r_{b'e}} \times \frac{1}{1+\dfrac{j\omega C_{in}r_{b'e}r_{bb'}}{r_{bb'}+r_{b'e}}}$$

or

$$A_{v(hf)} = \frac{A_{v(mf)}}{1+j\omega\tau} \tag{1.26}$$

where $\tau = C_{in}r_{b'e}r_{bb'}/(r_{bb'}+r_{b'e})$.

It is clear from equation (1.26) that the high-frequency voltage gain of a transistor falls with increase in frequency and has fallen by 3 dB from its low-frequency value at the frequency at which $\omega\tau$ is unity.

Example 1.7

A common-emitter amplifier uses a bipolar transistor having the following data.

$$r_{bb'} = 80\ \Omega \qquad C_{b'e} = 100\ \text{pF} \qquad r_{b'e} = 1200\ \Omega$$

$$r_{b'c} = 2.5\ \text{M}\Omega \qquad C_{b'c} = 2\ \text{pF} \qquad r_{ce} = 60\ \text{k}\Omega$$

If $g_m = 38$ mS determine the value of purely resistive load for which the circuit has a 3 dB bandwidth of 3 MHz.

Solution
From equation (1.26)

$$\left|\frac{A_{v(hf)}}{A_{v(mf)}}\right| = \frac{1}{\sqrt{2}} = \frac{1}{\sqrt{[1+\omega_{3dB}^2\tau^2]}}$$

or $\tau = 1/\omega_{3dB}$.

Hence,

$$\frac{C_{in}r_{b'e}r_{bb'}}{r_{bb'}+r_{b'e}} = \frac{1}{2\pi \times 3 \times 10^6} = 53.052 \times 10^{-9}$$

$$C_{in} = \frac{53.052 \times 10^{-9} \times (80 + 1200)}{80 \times 1200} = 707.36 \text{ pF.}$$

Therefore

$$707.36 \times 10^{-12} = 100 \times 10^{-12} + 2 \times 10^{-12}(1 + 38 \times 10^{-3} R_L)$$

$$\frac{607.36 \times 10^{-12}}{2 \times 10^{-12}} = 1 + 38 \times 10^{-3} R_L$$

$$R_L = \frac{302.68}{38 \times 10^{-3}} = 7965 \ \Omega. \quad (Ans.)$$

High-frequency Current Gain

The short-circuit current gain of a bipolar transistor falls at the higher frequencies because of its inherent self-capacitances. The meanings of the terms *cut-off* and *transition* frequencies have already been explained (page 3) and the hybrid-π equivalent circuit will now be used to derive expressions for these frequencies.

Consider the simplified equivalent circuit given in Fig. 1.19 again. To obtain an expression for the short-circuit current gain h_{fe} of the transistor, set R_L to zero. Then, from equation (1.24)

$$C_{in} = C_{b'e} + C_{b'c}.$$

Now

$$I_b = V_{b'e}\left[\frac{1}{r_{b'e}} + j\omega(C_{b'e} + C_{b'c})\right] \quad \text{and} \quad I_c = g_m V_{b'e}.$$

Therefore,

$$h_{fe} = I_c/I_b = \frac{g_m}{\dfrac{1}{r_{b'e}} + j\omega(C_{b'e} + C_{b'c})}$$

$$= \frac{g_m r_{b'e}}{1 + j\omega r_{b'e}(C_{b'e} + C_{b'c})}.$$

The transition frequency f_t is the frequency at which the magnitude of h_{fe} has fallen to unity. Hence,

$$1 \simeq g_m/\omega_t(C_{b'e} + C_{b'c})$$

$$\text{or} \quad f_t = \frac{g_m}{2\pi(C_{b'e} + C_{b'c})}. \tag{1.27}$$

h_{fe} has fallen by 3 dB from its low-frequency value $h_{fe(lf)}$ at the *cut-off frequency* f_β.

$$\left|\frac{h_{fe}}{h_{fe(lf)}}\right| = \frac{1}{\sqrt{2}} = \frac{1}{\sqrt{[1 + \omega_\beta^2(C_{b'e} + C_{b'c})^2 r_{b'e}^2]}}$$

Thus, $1 = \omega_\beta(C_{b'e}+C_{b'c})r_{b'e}$ and

$$f_\beta = \frac{1}{2\pi(C_{b'e}+C_{b'c})r_{b'e}}. \tag{1.28}$$

Multiplying equation (1.28) by the low-frequency value of h_{fe}, i.e. $g_m r_{b'e}$ gives

$$h_{fe(lf)} \times f_\beta = f_t$$

and this confirms that f_t is the short-circuit gain-bandwidth product of the transistor and thus equation (1.3) can be used to find h_{fe} at any frequency.

y-parameter Equivalent Circuit

The use of the h-parameter equivalent circuit is restricted to the lower frequencies at which the effects of transistor capacitances can be neglected. At higher frequencies the hybrid-π circuit can be used but also available is the y-parameter equivalent circuit. There are two reasons for the use of y-parameters at higher frequencies.

1 The y-parameters are easily and directly measured.
2 They are well suited to nodal analysis.

Referring to Fig. 1.13 the a.c. performance of a transistor can be described by the equations

$$I_b = y_{ie}V_{be}+y_{re}V_{ce} \tag{1.29}$$

$$I_c = y_{fe}V_{be}+y_{oe}V_{ce}. \tag{1.30}$$

It will be noted that the input and output currents are expressed in terms of the input and output voltages; this means that all four parameters have the dimensions of *admittance*.

$y_{ie} = I_b/V_{be}$ with $V_{ce}=0$ and is the short-circuit input admittance.
$y_{re} = I_b/V_{ce}$ with $V_{be}=0$ and is the short-circuit reverse-transfer admittance.
$y_{fe} = I_c/V_{be}$ with $V_{ce}=0$ and is the short-circuit forward-transfer admittance.
$y_{oe} = I_c/V_{ce}$ with $V_{be}=0$ and is the short-circuit output admittance.

The a.c. equivalent circuit described by these equations is shown by Fig. 1.20.

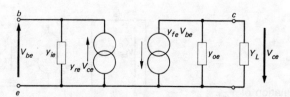

Fig. 1.20 *y*-parameter equivalent circuit of a transistor.

Voltage and Current Gain

The voltage gain $A_v = V_{ce}/V_{be}$ of a transistor is determined by writing

$$V_{ce} = -I_c R_L = -I_c/Y_L$$

and substituting into equation (1.30).

$$-V_{ce}Y_L = y_{fe}V_{be} + y_{oe}V_{ce}$$

$$-V_{ce}(Y_L + y_{oe}) = y_{fe}V_{be}.$$

Therefore,

$$A_v = V_{ce}/V_{be} = -y_{fe}/(Y_L + y_{oe}). \qquad (1.31)$$

Input and Output Admittances

The input admittance is $Y_{in} = I_b V_{be}$. From equation (1.30),

$$-V_{ce}Y_L = y_{fe}V_{be} + y_{oe}V_{ce}$$

$$V_{ce} = -V_{be}y_{fe}/(Y_L + y_{oe})$$

Substituting into equation (1.29)

$$I_b = y_{ie}V_{be} - y_{re}y_{fe}V_{be}/(Y_L + y_{oe})$$

and so

$$Y_{in} = I_b/V_{be} = y_{ie} - \frac{y_{re}y_{fe}}{Y_L + y_{oe}} \qquad (1.32)$$

The output admittance of the transistor is $Y_{out} = I_c/V_{ce}$ siemen. From Fig. 1.21 $V_{be} = -I_b R_s = -I_b/Y_s$. Substituting into equation (1.29)

$$-Y_s V_{be} = y_{ie}V_{be} + y_{re}V_{ce}$$

$$-V_{be}(Y_s + y_{ie}) = y_{re}V_{ce}$$

$$V_{be} = -\frac{y_{re}V_{ce}}{Y_s + y_{ie}}.$$

Substituting for V_{be} into equation (1.30) gives

$$I_c = y_{oe}V_{ce} - \frac{y_{fe}y_{re}V_{ce}}{Y_s + y_{ie}}$$

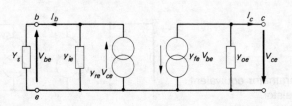

Fig. 1.21 Calculation of Y_{out}.

and

$$Y_{out} = I_c/V_{ce} = y_{oe} - \frac{y_{fe}y_{re}}{Y_s + y_{ie}}. \tag{1.33}$$

Current Gain

Rearranging the relationship $A_v = A_i R_L/R_{in}$ gives

$$A_i = A_v R_{in}/R_L = A_v Y_L/Y_{in}.$$

Hence,

$$A_i = I_c/I_b = \frac{-y_{fe}Y_L}{(Y_L + y_{oe})\left(y_{ie} - \dfrac{y_{re}y_{fe}}{Y_L + y_{oe}}\right)}$$

$$= \frac{-y_{fe}Y_L}{y_{ie}Y_L + y_{ie}y_{oe} - y_{re}y_{fe}}. \tag{1.34}$$

Example 1.8

A common-emitter transistor amplifier has a collector load resistor of 6.8 kΩ. The y-parameters of the transistor are

$$y_{ie} = 480 \times 10^{-6} \text{ S} \qquad y_{fe} = 0.04 \text{ S}$$

$$y_{oe} = 40 \times 10^{-6} \text{ S} \qquad y_{re} = -2 \times 10^{-6} \text{ S}.$$

If the source resistance is 500 Ω calculate (a) the input resistance, (b) the output resistance, and (c) the voltage gain.

Solution

(a) From equation (1.32)

$$Y_{in} = 480 \times 10^{-6} + \frac{0.04 \times 2 \times 10^{-6}}{\dfrac{1}{6.8 \times 10^3} + 40 \times 10^{-6}} = 907.7 \times 10^{-6} \text{ S}$$

and $R_{in} = 1/Y_{in} = 1102 \ \Omega$. (Ans.)

(b) From equation (1.33)

$$Y_{out} = 40 \times 10^{-6} + \frac{0.04 \times 2 \times 10^{-6}}{\dfrac{1}{500} + 480 \times 10^{-6}} = 72.26 \times 10^{-6} \text{ S}$$

and $R_{out} = 1/Y_{out} = 13\ 839 \ \Omega$. (Ans.)

(c) From equation (1.31)

$$A_v = \frac{0.04}{\dfrac{1}{6.8 \times 10^3} + 40 \times 10^{-6}} = 213.8. \quad (Ans.)$$

Field-effect Transistor Equivalent Circuit

The equivalent circuit of a field-effect transistor is shown in Fig. 1.22(a) in which

$$g_m = I_d/V_{gs} \, S \qquad (V_{DS} \text{ constant})$$

$$r_{ds} = V_{ds}/I_d \, \Omega \qquad (V_{GS} \text{ constant}).$$

At high frequencies the internal capacitances of a FET can no longer be neglected and the equivalent circuit that must be used is given by Fig. 1.22(b). Typical values for these components are $C_{gs} = 4$ pF, $C_{gd} = 1$ pF, and $r_{ds} = 100$ kΩ.

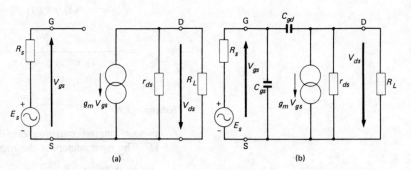

Fig. 1.22 Equivalent circuit of a FET at:
(a) low and medium frequencies,
(b) high frequency.

At low frequencies the voltage gain of a FET is simply

$$A_v = V_{ds}/V_{gs} = g_m V_{gs} R_L/V_{gs} = g_m R_L \qquad (1.35)$$

(assuming that $r_{ds} \gg R_L$).

As the frequency is increased, the voltage gain $V_{ds}/V_{gs} = V_{ds}/E_s$ falls because of the effect of the internal capacitances C_{gs} and C_{gd}. The input capacitance C_{in} of the FET is

$$C_{in} = C_{gs} + C_{gd}(1 + g_m R_L)$$

and hence

$$V_{gs} = \frac{E_s \times 1/j\omega C_{in}}{R_s + 1/j\omega C_{in}} = \frac{E_s}{1 + j\omega C_{in} R_s}$$

and

$$V_{ds} = \frac{g_m R_L E_s}{1 + j\omega C_{in} R_s}$$

so that the voltage gain is

$$A_v = V_{ds}/E_s = \frac{g_m R_L}{1 + j\omega C_{in} R_s}. \qquad (1.36)$$

This analysis neglects the capacitance C_{ds} between the drain and the source but the error involved is small since the input time constant is predominant.

The voltage gain falls by 3 dB from its low-frequency value at the frequency f_{3dB} which makes $1 = \omega C_{in} R_s$ or

$$f_{3dB} = 1/2\pi C_{in} R_s. \tag{1.37}$$

The gain-bandwidth product is

$$A_{v(lf)} \times f_{3dB} = \frac{g_m R_L}{2\pi C_{in} R_s}. \tag{1.38}$$

Example 1.9

A FET amplifier has a high-frequency 3 dB point of 30 kHz. Calculate the frequency at which the gain has fallen by 10 dB from its low-frequency value.

Solution
From equation (1.36)

$$\left| \frac{A_v}{A_{v(lf)}} \right| = \frac{1}{\sqrt{10}} = \frac{1}{\sqrt{[1 + \omega_{10dB}^2 C_{in}^2 R_s^2]}}$$

and $3 = \omega_{10dB} C_{in} R_s$ or

$$f_{10dB} = \frac{3}{2\pi C_{in} R_s} = 3 \times 30 \text{ kHz} = 90 \text{ kHz}. \quad (Ans.)$$

Data Sheets

The selection of a transistor for a particular application can best be made by consulting the data sheets provided by the manufacturer. A great deal of information about a device is provided by a data sheet and to make full use of it, it is essential to understand the meanings of the various symbols which are employed. The parameters of a particular type of transistor vary considerably from one device to another and a data sheet gives the typical value, very often with maximum and/or minimum values also quoted. The current gain of a bipolar transistor depends upon the collector current and so the collector current at which the maximum current gain is obtained is given. The minimum value of a breakdown voltage is normally given.

In addition to giving the absolute maximum ratings of a device a data sheet usually includes details of typical *h*- or *y*-parameters and provides graphs of such things as output or drain characteristics, input and transfer characteristics, and the variation of *h*- or *y*-parameters with collector or drain current or voltage, or with frequency.

Information may also be provided about the noise performance, the power dissipation derating and the high-frequency characteristics.

A.F. SILICON PLANAR EPITAXIAL TRANSISTORS

N-P-N transistors in TO-18 metal envelopes with the collector connected to the case.

The **BC107** is primarily intended for use in driver stages of audio amplifiers and in signal processing circuits of television receivers.

The **BC108** is suitable for multitude of low-voltage applications e.g. driver stages or audio preamplifiers and in signal processing circuits of television receivers.

The **BC109** is primarily intended for low-noise input stages in tape recorders, hi-fi amplifiers and other audio-frequency equipment.

QUICK REFERENCE DATA

			BC107	BC108	BC109	
Collector-emitter voltage ($V_{BE} = 0$)	V_{CES}	max.	50	30	30	V
Collector-emitter voltage (open base)	V_{CEO}	max.	45	20	20	V
Collector current (peak value)	I_{CM}	max.	200	200	200	mA
Total power dissipation up to T_{amb} = 25 °C	P_{tot}	max.	300	300	300	mW
Junction temperature	T_j	max.	175	175	175	°C
Small-signal current gain at T_j = 25 °C I_C = 2 mA; V_{CE} = 5 V; f = 1 kHz	h_{fe}	> <	125 500	125 900	240 900	
Transition frequency at f = 35 MHz I_C = 10 mA; V_{CE} = 5 V	f_T	typ.	300	300	300	MHz
Noise figure at R_S = 2 kΩ I_C = 200 μA; V_{CE} = 5 V f = 30 Hz to 15 kHz	F	typ. <	– –	– –	1,4 4,0	dB dB
f = 1 kHz; B = 200 Hz	F	typ.	2	2	1,2	dB

RATINGS Limiting values in accordance with the Absolute Maximum System (IEC 134)

Voltages

			BC107	BC108	BC109	
Collector-base voltage (open emitter)	V_{CBO}	max.	50	30	30	V
Collector-emitter voltage ($V_{BE} = 0$)	V_{CES}	max.	50	30	30	V
Collector-emitter voltage (open base)	V_{CEO}	max.	45	20	20	V
Emitter-base voltage (open collector)	V_{EBO}	max.	6	5	5	V

Currents

Collector current (d.c.)	I_C	max.	100	mA
Collector current (peak value)	I_{CM}	max.	200	mA
Emitter current (peak value)	$-I_{EM}$	max.	200	mA
Base current (peak value)	I_{BM}	max.	200	mA

Power dissipation

Total power dissipation up to T_{amb} = 25 °C	P_{tot}	max.	300	mW

Temperatures

Storage temperature	T_{stg}		–65 to +175	°C
Junction temperature	T_j	max.	175	°C

THERMAL RESISTANCE

From junction to ambient in free air	$R_{th\ j-a}$	=	0.5	°C/mW
From junction to case	$R_{th\ j-c}$	=	0.2	°C/mW

CHARACTERISTICS

T_j = 25 °C unless otherwise specified

Collector cut-off current

I_E = 0; V_{CB} = 20 V; T_j = 150 °C	I_{CBO}	<	15	μA

Base-emitter voltage [1]

I_C = 2 mA; V_{CE} = 5 V	V_{BE}	typ. 550 to	620 700	mV mV
I_C = 10 mA; V_{CE} = 5 V	V_{BE}	<	770	mV

[1] V_{BE} decreases by about 2 mV/°C with increasing temperature.

Fig. 1.23 Data sheet for BC 107/108/109 bipolar transistors (*Courtesy Philips Components Ltd*).

CHARACTERISTICS (continued) T_j = 25 °C unless otherwise specified

Saturation voltages [2])

I_C = 10 mA; I_B = 0.5 mA

V_{CEsat}	typ.	90	mV
	<	250	mV
V_{BEsat}	typ.	700	mV

I_C = 100 mA; I_B = 5 mA

V_{CEsat}	typ.	200	mV
	<	600	mV
V_{BEsat}	typ.	900	mV

Knee voltage

I_C = 10 mA; I_B = value for which
I_C = 11 mA at V_{CE} = 1 V

V_{CEK}	typ.	300	mV
	<	600	mV

Collector capacitance at f = 1 MHz

I_E = I_e = 0; V_{CB} = 10 V

C_c	typ.	2.5	pF
	<	4.5	pF

Emitter capacitance at f = 1 MHz

I_C = I_c = 0; V_{EB} = 0.5 V

C_e	typ.	9	pF

Transition frequency at f = 35 MHz

I_C = 10 mA; V_{CE} = 5 V

f_T	typ.	300	MHz

Small signal current gain at f = 1 kHz

			BC107	BC108	BC109	
I_C = 2 mA; V_{CE} = 5 V	h_{fe}	>	125	125	240	
		<	500	900	900	

Noise figure at R_S = 2 kΩ
I_C = 200 μA; V_{CE} = 5 V

			BC107	BC108	BC109	
f = 30 Hz to 15 kHz	F	typ.			1.4	dB
		<			4	dB
f = 1 kHz; B = 200 Hz	F	typ.	2	2	1.2	dB
		<	10	10	4	dB

			BC107A BC108A	BC107B BC108B BC109B	BC108C BC109C	
D.C. current gain						
I_C = 10 μA; V_{CE} = 5 V	h_{FE}	>		40	100	
		typ.	90	150	270	
I_C = 2 mA; V_{CE} = 5 V	h_{FE}	>	110	200	420	
		typ.	180	290	520	
		<	220	450	800	
h parameters at f = 1 kHz (common emitter)						
I_C = 2 mA; V_{CE} = 5 V						
Input impedance	h_{ie}	>	1.6	3.2	6	kΩ
		typ.	2.7	4.5	8.7	kΩ
		<	4.5	8.5	15	kΩ
Reverse voltage transfer ratio	h_{re}	typ.	1.5	2	3	10^{-4}
Small signal current gain	h_{fe}	>	125	240	450	
		typ.	220	330	600	
		<	260	500	900	
Output admittance	h_{oe}	typ.	18	30	60	$\mu\Omega^{-1}$
		<	30	60	110	$\mu\Omega^{-1}$

Fig. 1.23 cont.

[2]) V_{BEsat} decreases by about 1.7 mV/°C with increasing temperature.

Typical behaviour of collector current versus collector-emitter voltage

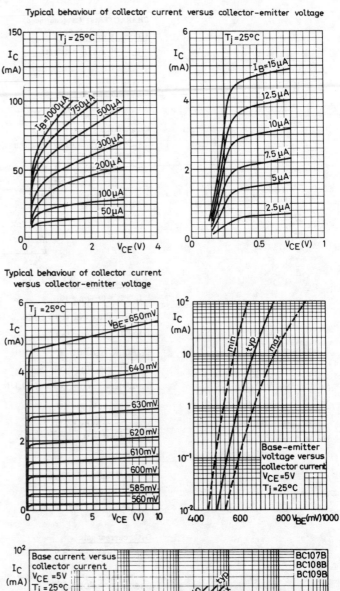

Typical behaviour of collector current versus collector-emitter voltage

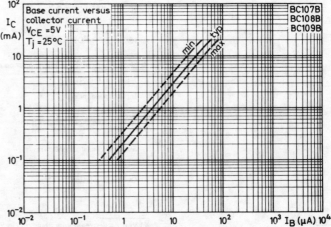

Fig. 1.23 cont.

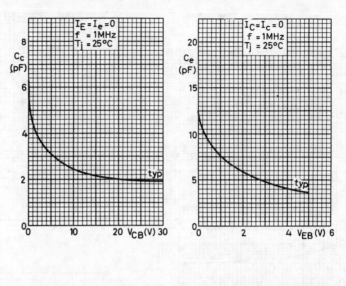

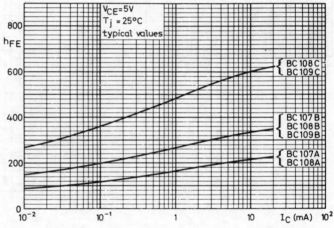

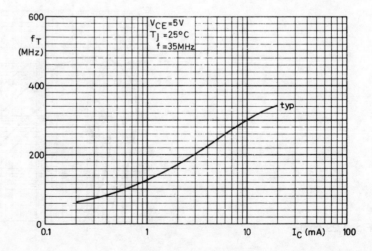

Fig. 1.23 cont.

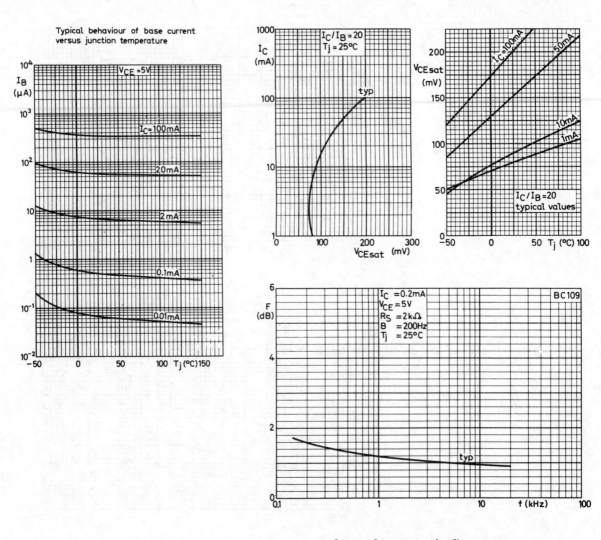

Typical behaviour of base current versus junction temperature

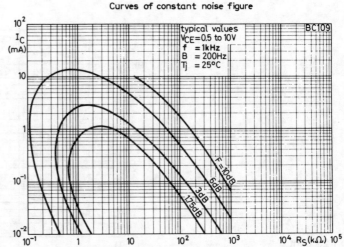

Curves of constant noise figure

Fig. 1.23 cont.

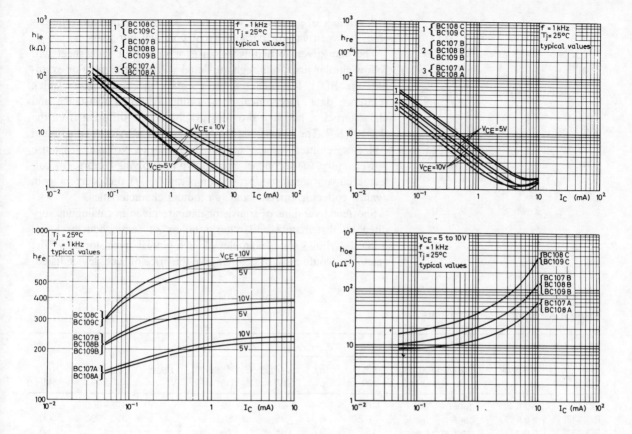

Fig. 1.23 cont.

The data sheet for the BC 107, 108 and 109 n-p-n transistors is given in Fig. 1.23. The data sheet starts by listing the intended main applications of the three transistors and follows with quick reference data. This is provided to allow a simple comparison to be made between these transistors and other types whose possible use is under consideration. Then follows mechanical data which shows the package that is employed.

Next, the absolute maximum ratings at 25 °C gives figures for the limiting values of voltage, current, power and temperature for the devices. These quoted values should never be exceeded when the transistor is connected in a circuit, otherwise failure of the device is likely. For better reliability it is good practice to ensure that an adequate safety margin is applied. If the device is to be operated at a temperature in excess of 25 °C all the quoted ratings should be downgraded.

After giving the thermal resistance of the transistors the data sheet lists the characteristics of the transistors, also at 25 °C. These specify the current and voltage values at which the quoted figures are applicable. Some adjustment will be necessary for operation at some other current, voltage or temperature values.

A number of graphs are then given which illustrate the relationships between various parameters of potential importance.

The data given in a power-transistor or a FET data sheet will follow similar lines, although the detail will be different. For example, the BD 131 power transistor's data sheet also gives quick reference data, mechanical data, ratings, thermal resistance and characteristics, but then provides graphical data different from the BC 107/8/9. The graphs that are given are (*a*) SOAR, (*b*) maximum collector—emitter voltage against base—emitter voltage, (*c*) pulse power rating, (*d*) secondary breakdown multiplying factors, (*e*) collector current against junction temperature, (*f*) d.c. current gain against collector current, and (*g*) mutual characteristics.

Shortened versions of transistor data are given in catalogues supplied by distributors and other retailers of semiconductors, and Table 1.4 shows some examples of such bipolar transistor data. The catalogue would also, of course, give the price of each device listed.

Table 1.4 Bipolar transistor data

| Type | Material | Application | Maximum | | | | | h_{FE} | f_t |
			P_D (W)	I_C (A)	V_{CEO} (V)	V_{CBO} (V)	V_{EBO} (V)		(MHz)
BC 108	n-p-n Si	general	0.30	0.1	20	30	5	125–900	300
BC 547	n-p-n Si	general	0.50	0.1	45	50	6	125–900	300
BD 131	n-p-n Si	medium power	15	3	45	70	6	40	60

In choosing a transistor for a particular purpose, the first step should be to decide the general type which is required, e.g. audio- or radio-frequency, low-, medium- or high-power, low-noise or switching. Very often a general-purpose device will suffice.

For a high-frequency application a transistor whose f_t is much higher than the highest frequency to be handled should be selected and then specific choice made on the grounds of voltage and current rating and power dissipation (which may not be of concern for a small-signal amplifier), and the current gain. For audio-frequencies, the f_t figure is not usually an important factor as long as it is high enough for adequate current gain at the highest frequency of concern (see eqn 1.3). Then take into account the wanted ratings of the device. For high-power applications the merits of the power MOSFET should also be considered.

For the less-demanding applications it will probably be found that several transistors prove to be equally suitable; the choice between them can then be made on the grounds of cost, availability, and second-sourcing.

2 Small-signal Audio-frequency Amplifiers

The function of a small-signal audio-frequency amplifier is to deliver a current or a voltage to a load, the power output being relatively unimportant. The important characteristics of a small-signal amplifier are

(*a*) its voltage or current gain;
(*b*) its maximum output signal voltage or current; and
(*c*) its input and output impedances.

Very often its noise performance is of importance too. Sometimes it is necessary to apply some degree of *negative feedback* to the circuit in order to meet all of its specifications but this topic will be deferred until the next chapter. The active devices used to provide the required amplification may be either a transistor or an integrated circuit. Generally, small-signal integrated amplifiers use one of the many kinds of operational amplifier which are available, but for low-noise applications it may be necessary to use a low-noise a.f. preamplifier.

The term *small-signal* implies that the swings of collector or drain current or voltage are small enough for the parameters of the device to be considered as essentially constant. This means that the determination of the gain, etc. can be carried out using one of the a.c. equivalent circuits discussed in Chapter 1.

Single-stage Transistor Amplifiers

Bias Circuit

The *operating point* of a bipolar transistor must be chosen to satisfy one or more of the following factors.

1 Transistors of the same type, even when made in the same batch, very often possess widely differing h_{FE} values, commonly over a 1−4 or even greater range.
2 The values of h_{FE}, I_{CEO} and V_{BE} are all temperature dependent.
3 The current gain depends upon the collector current and the maximum gain is obtained at a collector current specified by the manufacturer.
4 For maximum output voltage, the operating point must be at the middle of the d.c. load line so that the maximum possible swings of current and voltage can be obtained.

5 The input resistance is a function of the collector current and for a high impedance a low value of current is necessary.

6 The transition frequency f_t also depends upon the magnitude of the collector current.

7 As the collector current is reduced so the unwanted noise voltages generated within a transistor reduce in magnitude.

The selected operating point must be specified by applying a *bias* current to the transistor. An amplifier stage can be designed to have the chosen d.c. collector current assuming the *nominal* value of h_{FE} for the transistor. The bias circuit should operate to ensure that approximately the same collector current will flow if a transistor having either the maximum or the minimum h_{FE} should be used instead. This means that the bias circuit must also provide *d.c. stabilization* to keep the collector current at more or less the chosen value, even though h_{FE}, I_{CEO} and V_{BE} may vary considerably.

The most commonly employed bias circuit is the potential divider arrangement shown in Fig. 2.1. The circuit operates to provide d.c. stabilization in the following way. The base–emitter voltage V_{BE} of the transistor is equal to the voltage $V_B = V_{CC}R_2/(R_1+R_2)$ dropped across R_2 minus the voltage $V_E = I_E R_4$ across R_4. If the collector current should increase for *any* reason, the voltage across R_4 will rise and in so doing will reduce the forward-bias voltage V_{BE} of T_1. This will reduce the collector current and so tend to oppose the initial increase in the current. Signal-frequency currents flowing in the emitter resistance will apply *negative feedback* to the circuit. If this is not required, an emitter decoupling capacitor must be provided.

For analytical purposes the circuit shown in Fig. 2.1 can be simplified by the application of Thevenin's theorem to the left of the base terminal of T_1. The Thevenin equivalent circuit thus obtained is shown in Fig. 2.2 in which

$$V_B = V_{CC}R_2/(R_1+R_2) \quad \text{and} \quad R_B = R_1R_2/(R_1+R_2).$$

Hence, applying Kirchhoff's second law

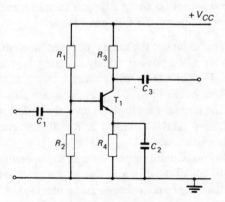

Fig. 2.1 Potential divider bias.

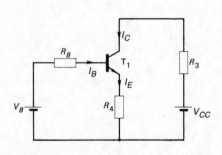

Fig. 2.2 Thevenin equivalent circuit of the bias circuit.

$$V_B = I_B R_B + V_{BE} + I_E R_4$$

$$V_B - V_{BE} = I_B (R_B + R_4) + I_C R_4$$

But $I_C = h_{FE} I_B + I_{CEO}$, therefore

$$V_B - V_{BE} = \frac{(I_C - I_{CEO})}{h_{FE}} (R_B + R_4) + I_C R_4$$

$$= \frac{I_C}{h_{FE}} (R_B + R_4 [1 + h_{FE}]) - \frac{I_{CEO}}{h_{FE}} (R_B + R_4)$$

$$I_C = \frac{(V_B - V_{BE}) h_{FE}}{R_B + R_4 (1 + h_{FE})} + \frac{I_{CEO} (R_B + R_4)}{R_B + R_4 (1 + h_{FE})} \qquad (2.1)$$

The stability of the operating point with respect to changes in one or more of the current gain h_{FE}, V_{BE} and I_{CEO} can be determined by differentiating equation (2.1). Note, however, that it is customary to consider the change in collector current caused by a change in I_{CBO} and *not* in I_{CEO}. Further, if a silicon device is used so that I_{CEO} is negligibly small and the component values are such that $h_{FE} R_4 \gg R_B$, then equation (2.1) reduces to $I_C = (V_B - V_{BE})/R_4$.

Provided $V_B \gg V_{BE}$, the collector current should remain sensibly constant even though h_{FE} should exhibit considerable variation.

For a germanium transistor the effect of changes in the collector leakage current is much larger and must often be taken account of, and a *stability function S* has been defined where

$$S = \delta I_C / \delta I_{CBO}. \qquad (2.2)$$

The second term of equation (2.1) can be rewritten as

$$\frac{I_{CBO} (1 + h_{FE}) (R_4 + R_B)}{R_B + R_4 (1 + h_{FE})}$$

and differentiating this with respect to I_{CBO} gives

$$S = \frac{(1 + h_{FE}) (R_4 + R_B)}{R_B + R_4 (1 + h_{FE})}. \qquad (2.3)$$

Ideally, S should be zero but, in practice, it is made as small as possible. For this, R_4 should be as large, and R_B as small, as possible. Unfortunately, increasing R_4 also increases the d.c. voltage developed across the emitter resistance and thereby limits the maximum possible output voltage. Also, reducing the value of R_B increases the shunting effect that the bias resistors have upon the signal path. As a result, the design of a bias circuit must be a compromise between conflicting factors.

Example 2.1

A single-stage transistor amplifier of the type shown in Fig. 2.1 has the following component values: $R_1 = 12$ kΩ, $R_2 = 6.8$ kΩ, $h_{FE} = 140$,

$R_4 = 1000\ \Omega$. Calculate its stability factor $\delta I_C / \delta I_{CBO}$ and determine the change in its collector current when I_{CBO} changes by 20 nA. Assume all other parameters are constant in value.

Solution

The total base resistance is

$$R_B = 12 \times 10^3 \times 6800/(12 \times 10^3 + 6800) = 4340\ \Omega.$$

From equation (2.3)

$$S = \frac{141(4340 + 1000)}{4340 + 141 \times 1000} = 5.18.\quad (Ans.)$$

Change in collector current $= 5.18 \times 20 = 104$ nA. (*Ans.*)

It is possible to similarly derive equations for the stability obtained when h_{FE} varies, but the main cause of h_{FE} variations is transistor substitution and then the change is usually too large for differentiation to yield accurate results. An alternative method is best used which is also valid when two or more of the three potential variables change simultaneously.

Suppose that both h_{FE} and I_{CBO} vary but V_{BE} remains constant. From Fig. 2.2,

$$V_B = I_B R_B + V_{BE} + (I_C + I_B) R_4$$

$$I_B = \frac{V_B - V_{BE} - I_C R_4}{R_B + R_4}$$

Suppose that h_{FE} varies from h_{FE1} to h_{FE2} while I_{CEO} changes from I_{CEO1} to I_{CEO2} causing I_C to alter from I_{C1} to I_{C2}. Then

$$I_{B1} = \frac{V_B - V_{BE} - I_{C1} R_4}{R_B + R_4}$$

and

$$I_{B2} = \frac{V_B - V_{BE} - I_{C2} R_4}{R_B + R_4}.$$

Therefore,

$$I_{B1} - I_{B2} = (I_{C2} - I_{C1}) R_4 / (R_B + R_4)$$

or

$$I_{C2} - I_{C1} = \frac{(I_{B1} - I_{B2})(R_B + R_4)}{R_4}. \tag{2.4}$$

Clearly the same result would be obtained if the changes in collector current were due only to a change in h_{FE}, probably the result of device replacement.

Example 2.2

A transistor has a current gain of $h_{FE} = 100$ and is used in a potential divider circuit in which $R_1 = 120$ kΩ and $R_2 = 18$ kΩ. The collector current is 1.2 mA at an ambient temperature of 20 °C. The collector leakage current I_{CBO} is 10 nA at 20 °C and doubles in value for every 10 °C rise in temperature. The current gain h_{FE} increases by 5% for every 10 °C rise in temperature. Calculate the emitter resistance which will ensure that the collector current will not exceed 1.22 mA at an ambient temperature of 50 °C. Assume V_{BE} is constant at 0.7 V.

Solution
At 20 °C, $h_{FE1}=100$, $I_{CEO1}=101 \times 10$ nA$=1.01$ μA, $I_{C1}=1.2$ mA.
At 50 °C, $h_{FE2} \simeq 116$, $I_{CBO2}=80$ nA so $I_{CEO2}=80$ nA $\times 117=9.36$ μA.
Now $I_C=h_{FE}I_B+I_{CEO}$ so that

$$1.2 \times 10^{-3} = 100I_{B1}+1.01 \times 10^{-6} \quad \text{or} \quad I_{B1} = 11.99 \ \mu\text{A}$$

$$1.22 \times 10^{-3} = 116I_{B2}+9.36 \times 10^{-6} \quad \text{or} \quad I_{B2} = 10.44 \ \mu\text{A}.$$

Also, $R_B = 120 \times 10^3 \times 18 \times 10^3/(120+18) \times 10^3 = 15.65$ kΩ.
From equation (2.4)

$$\frac{R_4}{R_4+15.65 \times 10^3} = \frac{(11.99-10.44) \times 10^{-6}}{1.22 \times 10^{-3}-1.2 \times 10^{-3}} = 7.75 \times 10^{-2}$$

$R_4 = 1315$ Ω. *(Ans.)*

This is the smallest value of R_4 which should be used and probably the nearest preferred value, i.e. 1500 Ω would be used.

Example 2.3

For the circuit shown in Fig. 2.3 calculate the collector current and the collector–emitter voltage if $V_{BE} = 0.6$ V, $h_{FE} = 120$ and $I_{CBO} = 10$ nA.

Solution
Applying Thevenin's theorem to the base circuit,

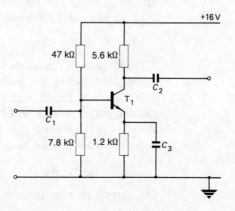

Fig. 2.3

$$V_B = \frac{16 \times 7.8}{7.8+47} = 2.28 \text{ V}$$

and the total base resistance is

$$R_B = \frac{7.8 \times 47}{7.8+47} = 6.69 \text{ k}\Omega.$$

Hence

$$2.28 = I_B \times 6.69 \times 10^3 + 0.6 + 1.2 \times 10^3 \times 121 I_B$$

$$I_B = \frac{1.68}{(6.69 + 1.2 \times 121) \times 10^3} = 11 \ \mu\text{A}.$$

Therefore,

$$I_C = 120 \times 11 \times 10^{-6} + 121 \times 10 \times 10^{-9} = 1.32 \text{ mA} \quad (Ans.)$$

$$V_{CE} = 16 - 1.32 \times 10^{-3} \times 5.6 \times 10^3 - 1.32 \times 10^{-3} \times 1.2 \times 10^3 = 7.0 \text{ V.}$$
$$(Ans.)$$

Design of the Bias Circuit

In the design of a potential divider bias circuit there are a greater number of variables than there are equations and so some element of judgement is necessary. This means that there are a number of different ways in which a bias circuit can be designed.

Applying Kirchhoff's second law to the collector−emitter circuit of Fig. 2.1,

$$V_{CC} = V_{CE} + I_C R_3 + (I_C + I_B) R_4$$

$$V_{CC} - V_{CE} \simeq I_C (R_3 + R_4).$$

The operating point of the transistor is specified by the choice made for $V_{CC} - V_{CE}$. For maximum output voltage the operating point should lie at the middle of the d.c. load line and then $V_{CE} = V_{CC}/2$. Choosing this point,

$$V_{CC}/2 = I_C (R_3 + R_4).$$

The value of I_C can be estimated, or obtained, from the output characteristics of the transistor. Then

$$R_3 + R_4 = V_{CC}/2 I_C.$$

The division of resistance between R_3 and R_4 must now be determined. The higher the value of R_4 the better will be the d.c. stabilization of the circuit but, on the other hand, the maximum possible output voltage will be reduced. A reasonable choice is to allow $V_{CC}/10$ volts to be developed across R_4, then $R_4 = V_{CC}/10 I_C \ \Omega$. The base voltage of the transistor is now

$$V_B = V_{BE} + I_C R_4.$$

The current flowing through R_2 must be several times larger than the base current I_B so that V_B remains sensibly constant as required for good d.c. stabilization. Choose I_{R2} to be n times the base current; the larger the value of n the better will be the d.c. stability but the lower will be the values of R_1 and R_2. Then

$$V_{CC} - V_B = (I_{R2} + I_B)R_1$$

$$R_1 = (V_{CC} - V_B)/(I_{R2} + I_B)$$

and $R_2 = V_B/I_{R2}$.

Example 2.4

Design a potential divider bias circuit for a transistor with $h_{FE} = 100$ and $V_{BE} = 0.7$ V if the collector current is to be 1.2 mA at an ambient temperature of 20 °C and the collector supply voltage is 20 V.

Solution

Choose the collector–emitter voltage to be $V_{CC}/2 = 10$ V. Then

$$R_3 + R_4 = 10/1.2 \times 10^{-3} = 8.33 \text{ k}\Omega.$$

If $V_{CC}/10$ or 2 V are to be dropped across R_4 then

$$R_4 = 2/1.2 \times 10^{-3} = 1667 \ \Omega = 1800 \ \Omega \text{ preferred value.}$$

Now $R_3 = 8333 - 1667 = 6667 \ \Omega = 6800 \ \Omega$ preferred value.
The base voltage is equal to

$$0.7 + 1.2 \times 10^{-3} \times 1800 = 2.86 \text{ V}$$

and the base current is

$$1.2 \times 10^{-3}/100 = 12 \ \mu\text{A}.$$

The current in R_2 should be at least 10 times greater than the base current, say 120 μA. Hence

$$R_1 = (20 - 2.86)/132 \times 10^{-6} \text{ k}\Omega = 129.8 = 120 \text{ k}\Omega \text{ preferred value.}$$

Lastly,

$$R_2 = 2.86/120 \times 10^{-6} = 23.8 \text{ k}\Omega = 22 \text{ k}\Omega \text{ preferred value.}$$

Notice the values of R_1 and R_2 obtained are nearly the values used in Example 2.2. The emitter resistance of 1800 Ω is higher than the minimum value necessary to keep the collector current within the limits quoted in that example; with $R_4 = 1800$ Ω the collector current will only increase from 1.2 mA to 1.2154 mA when the quoted increases in h_{FE} and I_{CBO} occur.

Voltage Gain at Mid-band Frequencies

Consider the single-stage bipolar transistor amplifier shown by Fig. 2.1. Capacitors C_1 and C_3 act as input and output coupling capacitors whose purpose is to prevent the d.c. conditions of the stage being upset by the source and load resistances.

At medium, or mid-band, frequencies, the reactances of all three capacitors are negligibly small and can be assumed to be zero. The emitter resistor R_4 is effectively short-circuited and so it does not appear in the a.c. equivalent circuit of the amplifier. Since the collector power supply has negligible internal resistance, the bias resistors R_1 and R_2 are effectively connected in parallel with one another across the base−emitter terminals of the transistor. Also, since C_3 has negligible reactance, the external load resistor appears in parallel with the collector resistor R_3.

The h-parameter equivalent circuit of the amplifier is given by Fig. 2.4 in which $R = R_1 R_2/(R_1 + R_2)$. In the figure, h_{re} has been neglected and is not shown. Very often h_{oe} can also be omitted without the introduction of undue error.

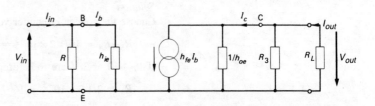

Fig. 2.4 h-parameter equivalent circuit of Fig. 2.1.

Inspection of the equivalent circuit makes it clear that the current gain of the amplifier will always be less than the current gain of the transistor itself. This is because some of the input current I_{in} is diverted away from the base of the transistor by the bias resistors R_1 and R_2 and because not all of the collector current flows into the load resistance R_L.

Example 2.5

Calculate the input resistance, the current gain and the voltage gain of the circuit shown in Fig. 2.5. The h-parameters of the transistor are $h_{ie} = 700\ \Omega$, $h_{oe} = 10^{-4}\ \text{S}$ and $h_{fe} = 250$, $(h_{re} \simeq 0)$. Determine also the percentage error in the calculation if h_{oe} is also neglected.

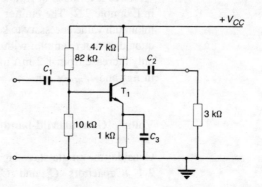

Fig. 2.5

Solution

The effective collector load resistance is 4.7 kΩ in parallel with 3 kΩ or 1.83 kΩ. Therefore, from equation (1.16) the current gain A_i of the transistor is

$$A_i = \frac{250}{1 + 10^{-4} \times 1.83 \times 10^3} = 211.3.$$

Since h_{re} is negligible, the input resistance of the transistor is equal to h_{ie} or 700 Ω. This value is reduced, by the shunting effect of the 82 kΩ and 10 kΩ resistors in parallel, to $700 \times 8913/(700 + 8913)$ or 649 Ω. Therefore, input resistance = 649 Ω. (*Ans.*)

The 8913 Ω effective resistance of the bias resistors reduces the base current I_b to

$$I_b = I_{in} \times 8913/(8913 + 700) = 0.927 I_{in}.$$

Similarly, at the output the current flowing in the load resistance R_L is

$$I_{out} = I_c \times 4.7/(4.7 + 3) = 0.61 I_c$$

Hence $I_{out} = 0.61 \times 211.3 \times 0.927 I_{in}$
and the overall current gain A_{io} is

$$A_{io} = I_{out}/I_{in} = 0.61 \times 211.3 \times 0.927 = 119.6. \quad (\textit{Ans.})$$

From equation (1.20) the voltage gain A_v of the circuit is

$$A_v = \frac{119.6 \times 3000}{649} = 552.9. \quad (\textit{Ans.})$$

If h_{oe} is neglected, the current gain of the transistor is equal to h_{fe}, i.e. 250. Then the overall current gain A_{io} is

$$A_{io} = I_{out}/I_{in} = 0.61 \times 250 \times 0.927 = 141.4. \quad (\textit{Ans.})$$

Hence,

$$\text{Voltage gain} = \frac{141.4 \times 3000}{649} = 653.6. \quad (\textit{Ans.})$$

The percentage error introduced by neglecting h_{oe} is

$$(653.6 - 552.9)/552.9 \times 100\% = +18.21\%. \quad (\textit{Ans.})$$

This error would be reduced if the effective load resistance were smaller. It should be borne in mind that the values of h_{ie}, h_{oe} and h_{fe} are not known accurately, since they vary between particular devices of the same type and it is doubtful whether there is much point in including h_{oe} in calculations.

The determination of gain and input—output impedances could, of course, also be carried out using either the hybrid-π or the y-parameter equivalent circuits.

Undecoupled Emitter Resistor

If the emitter resistor of the circuit of Fig. 2.1 is not decoupled, the voltage gain of the circuit will be reduced but will become less dependent upon the parameters of the device.

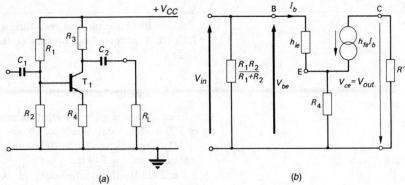

Fig. 2.6 (a) Amplifier with undecoupled emitter resistor, (b) equivalent circuit of (a).

Figure 2.6(a) shows a common-emitter amplifier with its emitter resistor undecoupled and Fig. 2.6(b) shows its h-parameter equivalent circuit. Both h_{re} and h_{oe} have been neglected and

$$R_L^l = R_3 R_L / (R_3 + R_L).$$

From Fig. 2.6(b) the current gain A_i of the transistor is

$$A_i = I_c / I_b = h_{fe} I_b / I_b = h_{fe}.$$

Thus the current gain is not affected by the undecoupled emitter resistor.

The input resistance R_{in} of the amplifier is equal to the input resistance of the transistor $R_{in(t)}$ in parallel with the bias components R_1 and R_2. From Fig. 2.6(b)

$$V_{be} = I_b h_{ie} + (1 + h_{fe}) I_b R_4.$$

Hence,

$$R_{in(t)} = V_{be} / I_b = h_{ie} + (1 + h_{fe}) R_4 \simeq h_{ie} + h_{fe} R_4. \qquad (2.5)$$

Usually, $h_{fe} R_4 \gg h_{ie}$ and then

$$R_{in(t)} = h_{fe} R_4. \qquad (2.6)$$

Clearly, one effect of the undecoupled emitter resistor is to increase the input resistance of the circuit although this is reduced by the bias resistors.

The voltage gain A_v of the circuit is

$$A_v = V_{out} / V_{in} = V_{ce} / V_{be} = \frac{I_c R_L^l}{I_b R_{in(t)}} = \frac{h_{fe} I_b R_L^l}{I_b (h_{ie} + h_{fe} R_4)}$$

$$= \frac{h_{fe} R_L^l}{h_{ie} + h_{fe} R_4}. \qquad (2.7)$$

Very often $h_{fe} R_4 \gg h_{ie}$ and then

$$A_v = \frac{h_{fe} R_L^l}{h_{fe} R_4} = \frac{R_L^l}{R_4}. \qquad (2.8)$$

The voltage gain given by equations (2.7) and (2.8) is considerably less than the value obtained with the emitter resistor decoupled but it is much more stable. The use of an undecoupled emitter resistor applies *negative feedback* to the circuit.

Emitter Follower

Figures 2.7(a) and (b) show, respectively, the circuit of an emitter follower and its a.c. equivalent circuit. When the input signal voltage goes positive, the base current is increased and the current gain of the transistor gives an increase in the emitter current. This current passes through the emitter load and develops a positive-going voltage, i.e. the output voltage *follows* the input voltage.

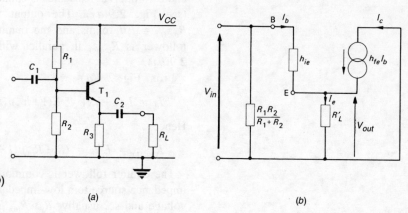

Fig. 2.7 (a) Emitter follower, (b) equivalent circuit of (a).

From Fig. 2.7(b) [in which $R_L^l = R_L R_3/(R_L+R_3)$] the current gain of the emitter follower circuit is

$$A_i = I_e/I_b = I_b(1+h_{fe})/I_b = 1+h_{fe}.$$

Also

$$V_{in} = I_b h_{ie} + (1+h_{fe})I_b R_L^l.$$

The input resistance $R_{in(t)}$ of the transistor is the ratio V_{in}/I_b and hence

$$R_{in(t)} = h_{ie} + (1+h_{fe})R_L^l \simeq h_{ie} + h_{fe}R_L^l. \tag{2.9}$$

Usually, $h_{fe}R_L^l \gg h_{ie}$ and then

$$R_{in(t)} \simeq h_{fe}R_L^l. \tag{2.10}$$

The input resistance of a transistor in the common-collector configuration is high but this value is reduced by the shunting effect of the bias resistors R_1 and R_2.

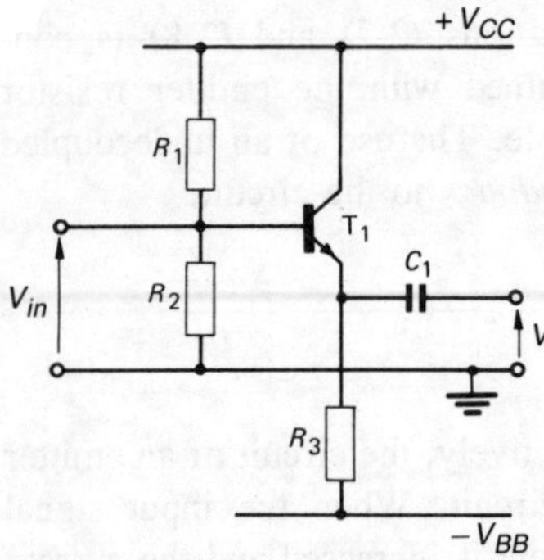

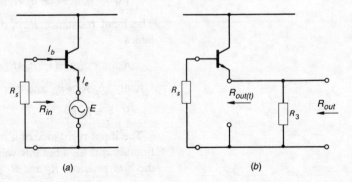

Fig. 2.8 Emitter follower with dual power supplies.

The voltage gain A_v of the circuit is the ratio V_{out}/V_{in} and so

$$A_v = \frac{I_b(1+h_{fe})R_L^l}{I_b(h_{ie}+h_{fe}R_L^l)} \simeq \frac{h_{fe}R_L^l}{h_{ie}+h_{fe}R_L^l}. \tag{2.11}$$

It should be clear from equation (2.11) that the voltage gain of an emitter follower must always be less than unity. The gain will more nearly approach unity as R_3 is increased in value but the maximum value that can be employed is determined by the collector supply voltage V_{CC} and the maximum output signal voltage that is required. The emitter resistance can be still further increased in value if a second power supply voltage is available, see Fig. 2.8.

The output resistance of an emitter follower can best be determined by assuming that the input terminals are closed in an impedance equal to the source impedance and that a generator of e.m.f. E volts and zero internal resistance is connected across the output terminals (see Fig. 2.9(a)). The output resistance of the transistor is $R_{out(t)} = E/I_e$ ohms and the output resistance R_{out} of the emitter follower is $R_{out(t)}$ in parallel with the emitter resistor R_3 (Fig. 2.9(b)).

From Fig. 2.9(a)

$$I_e = I_b(1+h_{fe}) = E(1+h_{fe})/(R_s+R_{in}).$$

Hence

$$R_{out(t)} = E/I_e = (R_s+R_{in})/(1+h_{fe}).$$

The emitter follower is commonly employed to connect a high-impedance source to a low-impedance load with little loss of signal voltage and so, usually, $R_s \gg R_{in}$. Then,

$$R_{out(t)} \simeq R_s/h_{fe} \tag{2.12}$$

and hence the output resistance of an emitter follower is

$$R_{out} = \frac{\dfrac{R_s}{h_{fe}} \times R_3}{(R_s/h_{fe})+R_3} = \frac{R_s R_3}{R_s+h_{fe}R_3}. \tag{2.13}$$

Fig. 2.9 Determination of the output resistance of an emitter follower.

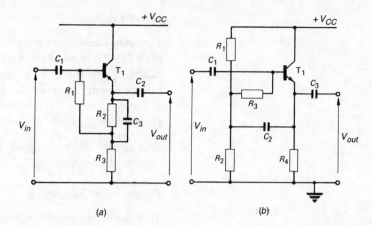

Fig. 2.10 Boot-strapped emitter followers.

(a)   (b)

The high input impedance of an emitter follower is shunted by the bias resistors R_1 and R_2. When a very high input impedance is needed, the circuit must be *bootstrapped* (see Fig. 2.10). In Fig. 2.10(a) the emitter resistance has been divided into two parts so that the required bias voltage appears at their junction. The junction of R_2 and R_3 is connected to the base of T_1 by resistor R_1. Since the voltage gain of an emitter follower is approximately unity, the signal voltages appearing at each end of R_1 are almost the same and so R_1 has an effective a.c. resistance that is very large. An alternative arrangement is shown by Fig. 2.10(b).

Multiple Stages

Very often the voltage gain wanted from an amplifier is larger than can be provided by a single stage. Then two or more stages must be connected in cascade. The overall voltage gain of a multi-stage amplifier is the product of the individual stage gains. For example, a two-stage amplifier has an overall voltage gain $A_v = A_1 A_2$ and the gain of an n-stage amplifier is $A_v = A_1 A_2 \ldots A_n$. The signal voltage is applied to the input terminals of the first stage, amplified, and then the amplified signal is applied to the input terminals of the next stage and so on for each stage in the amplifier. The output signal voltage of one stage is the input signal of the next stage and so some means of coupling the stages together is required. The overall current gain is not equal to the product of the current gains of the individual stages because of losses in the coupling circuits. There are several different ways in which inter-stage coupling can be accomplished, namely: capacitive, direct, optical, and transformer.

Capacitive Coupling

A capacitor is employed to connect the collector of one stage to the input of the next. *RC* coupling was commonly employed in discrete component designs but it is not suitable for use within an integrated circuit.

Direct Coupling

Two stages are direct coupled if the output of one is directly connected to the input of the next. The method introduces various difficulties (with drift and d.c. levels) if used with discrete circuitry but it is commonly used within integrated circuits.

Optical Coupling

Two circuits can be coupled, without any electrical connection, by the use of an opto-coupler.

Transformer Coupling

Transformer coupling is only used at audio frequencies to couple the output of some amplifiers to their load. Tuned transformer coupling is widely employed at radio frequencies where it not only couples stages together but also provides some measure of selectivity.

IC Amplifiers

Audio preamplifiers are designed to have a low noise figure, low distortion, wide bandwidth and good rejection of any power supply ripple. They are capable of providing a much better result than can an op-amp. Most IC preamplifiers are dual devices, i.e. two identical amplifiers within the one package, and the LM 381 is one of the most popular.

The data sheet for the LM 381 is given in Fig. 2.11. It can be seen that information is given on the package, absolute maximum ratings, and electrical characteristics, along with some graphical data.

The IC employs a differential amplifier input stage and direct inter-stage coupling. The input stage can be operated either as a single-, or as a differential-input amplifier, the former giving a better noise performance. The external biasing for the circuit differs according to whether a single-, or a differential-mode input is to be used.

Differential-mode Bias

The bias for the amplifier is set by the two external resistors R_1 and R_2 shown in Fig. 2.12(a) The non-inverting input is held at about $+1.2$ V above earth by internal circuitry, and the quiescent output voltage is set by the d.c. voltage across R_1. Usually, this is made to be one-half of the power supply voltage V_{CC} volts. For d.c. stability the current that flows in R_1 must be at least 5 μA and this means that the *maximum* value for R_1 is $1.2/(5 \times 10^{-6}) = 240$ kΩ. Also,

$$\left(\frac{R_1}{R_1+R_2}\right)\frac{V_{CC}}{2} = 1.2 \text{ V. Also, } R_2 = \left(\frac{V_{CC}}{2.4} - 1\right)R_1$$

National Semiconductor
LM381/LM381A Low Noise Dual Preamplifier

Audio/Radio Circuits

General Description

The LM381/LM381A is a dual preamplifier for the amplication of low level signals in applications requiring optimum noise performance. Each of the two amplifiers is completely independent, with individual internal power supply decoupler-regulator, providing 120 dB supply rejection and 60 dB channel separation. Other outstanding features include high gain (112 dB), large output voltage swing (V_{CC} − 2V) p-p, and wide power bandwidth (75 kHz, 20 V_{p-p}). The LM381/LM381A operates from a single supply across the wide range of 9 to 40 V.

Either differential input or single ended input configurations may be selected. The amplifier is internally compensated with the provision for additional external compensation for narrow band applications.

Features

- Low Noise — 0.5 μV total input noise
- High Gain — 112 dB open loop
- Single Supply Operation
- Wide supply range 9–40 V
- Power supply rejection 120 dB
- Large output voltage swing (V_{CC} − 2 V)$_{p-p}$
- Wide bandwidth 15 MHz unity gain
- Power bandwidth 75 kHz, 20 V_{p-p}
- Internally compensated
- Short circuit protected

Absolute Maximum Ratings

Supply voltage	+ 40 V
Power dissipation (Note 1)	715 mW
Operating temperature range	0 °C to 70 °C
Storage temperature range	− 65 °C to + 150 °C
Lead temperature (soldering, 10 s)	300 °C

Electrical Characteristics T_A = 25 °C, V_{CC} = 14 V, unless otherwise stated.

Parameter	Conditions	Min	Typ	Max	Units
Voltage gain	Open loop (differential input), f = 100 Hz		160,000		V/V
	Open loop (single ended), f = 100 Hz		320,000		V/V
Supply current	V_{CC} 9 to 40 V, R_L = ∞		10		mA
Input resistance					
(positive input)			100		kΩ
(negative input)			200		kΩ
Input current					
(negative input)			0.5		μA
Output resistance	Open loop		150		Ω
Output current	Source		8		mA
	Sink		2		mA
Output voltage swing	Peak-to-peak	V_{CC} − 2			V
Unity gain bandwidth			15		MHz
Power bandwidth	20 V_{p-p} (V_{CC} = 24 V)		75		kHz
Maximum input voltage	Linear operation			300	mVrms
Supply rejection ratio	f = 1 kHz		120		dB
Channel separation	f = 1 kHz		60		dB
Total harmonic distortion	60 dB Gain, f = 1 kHz		0.1		%
Total equivalent input noise	R_S = 600 Ω, 10 – 10,000 Hz (single-ended input, flat gain circuit, A_V = 100)				
LM381A			0.5	0.7	μVrms
LM381			0.5	1.0	μVrms

Note 1: For operation in ambient temperatures above 25 °C, the device must be derated based on a 150 °C maximum junction temperature and a thermal resistance of 175 °C/W junction to ambient.

Fig. 2.11 Data sheet for LM 381 low-noise dual pre-amplifier (*Courtesy National Semiconductor (UK) Ltd*).

Typical performance characteristics

Large signal frequency
response

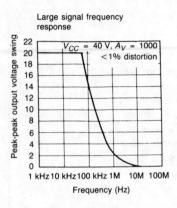

V_{CC} vs I_{CC}

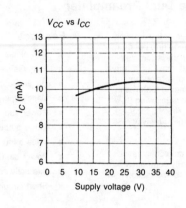

P-P Output voltage swing vs
V_{CC}

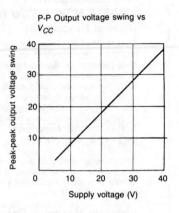

% Distortion

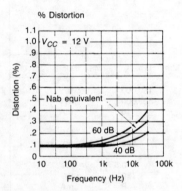

Channel Separation

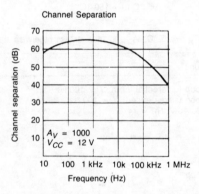

PSRR vs frequency

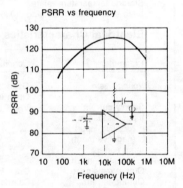

Gain and phase response

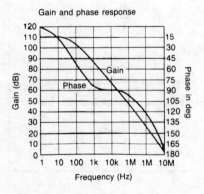

Noise voltage vs frequency

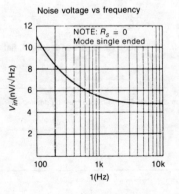

Noise current vs frequency

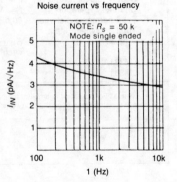

Fig. 2.11 cont.

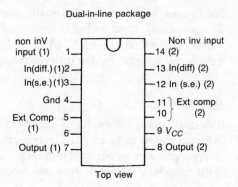

Fig. 2.11 cont.

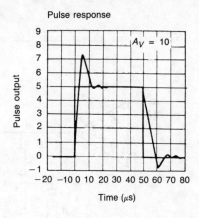

If the value of, say, R_1 is arbitrarily chosen, the value of R_2 can be calculated.

The closed-loop a.c. gain of the amplifier is given by $A_v = (R_1+R_2)/R_1$. Should the value thus obtained, using the values calculated to give the bias, be too low then R_1 can be shunted by a smaller resistor R_3 connected in series with a capacitor C_2, as shown by Fig. 2.12(b). Then the voltage gain $A_v = (R_2+R_3)/R_3$. The voltage gain of the amplifier will fall at lower frequencies because of the increasing reactance of C_2 and it will have fallen by 3 dB on its mid-band value at a frequency f_1, where

$$f_1 = 1/2\pi C_2 R_3 \text{ Hz.} \tag{2.14}$$

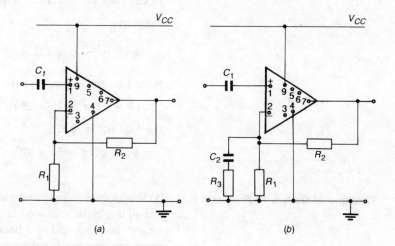

Fig. 2.12 The LM 381 audio-pre-amplifier with differential-mode bias.

Single-input Mode Bias

When a single-input circuit is required the bias arrangement must be slightly different, see Fig. 2.13. The inverting input terminal is connected to earth and the junction of the resistors R_1 and R_2 is connected to terminal 3. For adequate d.c. stability the maximum value

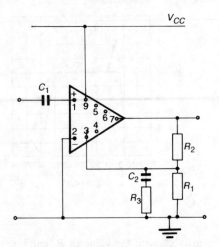

Fig. 2.13 The LM 381 audio pre-amplifier with single-input mode bias.

for R_1 must be only 1200 Ω. Usually, the circuit is biased to make the quiescent output voltage equal to $V_{CC}/2$ and since the d.c. voltage at terminal 3 is 0.6 V, $[R_1/(R_1+R_2)]$. $V_{CC}/2 = 0.6$. The a.c. gain is the same as for the differential-input circuit.

Bandwidth

The small-signal bandwidth of the LM381 is about 15 MHz; if this is higher than is required it can be reduced by connecting a capacitor of appropriate value between terminals 5 and 6. If the wanted upper 3 dB frequency is f_2 then the capacitance C_3 should be

$$C_3 = \frac{1}{2\pi f_2 A_v} - 4\,\mathrm{pF}. \tag{2.15}$$

Example 2.6

Design a LM 381 differential-input amplifier circuit to have a voltage gain of 30, a lower 3 dB point of 15 Hz, and an upper 3 db point of 30 kHz. The power supply voltage is 20 V.

Solution

$$\left(\frac{R_1}{R_1+R_2}\right)10 = 1.2, \quad \text{so that} \quad R_2 = 7.33R_1.$$

Choose a value for R_1 which is less than the maximum allowable, say 10 kΩ. Then $R_2 = 73.3$ kΩ. (*Ans.*)

Therefore,

$$30 = \frac{R_3+R_2}{R_3}, \quad \text{so that} \quad R_3 = 73.3/20 = 2.5 \text{ k}\Omega. \quad (\textit{Ans.})$$

For a 3 dB fall in gain at 15 Hz

$$C_2 = 1/(2\pi \times 15 \times 2500) \simeq 4 \ \mu\mathrm{F}. \quad (\textit{Ans.})$$

For 3 dB fall at 30 kHz

$$C_3 = 1/(2\pi \times 30 \times 10^3 \times 30) - 4 \times 10^{-12} = 177 \ \mathrm{nF}. \quad (\textit{Ans.})$$

Variation of Gain with Frequency

At low frequencies the reactances of any coupling capacitors and/or decoupling capacitors used in an a.f. amplifier increase and may no longer be regarded as being negligibly small. The coupling capacitors are connected in series with the signal path and some of the signal voltage will be dropped across these components. The increased reactance of the emitter decoupling capacitor allows some signal-frequency current to flow in the emitter resistor. Negative feedback is then applied to the amplifier and reduces the gain. For these two reasons the gain of an a.f. amplifier may fall at low frequencies. The gain of an amplifier will also fall at high frequencies.

There are two reasons for this: (*a*) various stray and transistor capacitances are in parallel with the signal path and at high frequencies their reactance becomes low enough to affect the gain; (*b*) the current gain of a transistor falls with increase in frequency.

Gain at Low Frequencies

The voltage gain of an audio-frequency amplifier will fall at low frequencies because of the time constants of the input and output coupling circuits. Figure 2.14 shows the input circuit of an amplifier which has an input resistance R_{in}. At middle frequencies when coupling capacitor C_1 has negligible reactance

$$V_{in} = V_{be} = E_s R_{in}/(R_s + R_{in})$$

and the output voltage of the amplifier is

$$V_{out} = g_m V_{be} R_L$$

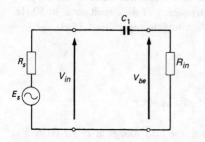

Fig. 2.14 Input circuit of an amplifier.

Hence the medium-frequency voltage gain V_{out}/V_{in} is

$$A_{v(mf)} = g_m R_L.$$

At low frequencies

$$V_{be} = \frac{E_s R_{in}}{R_s + R_{in} + 1/j\omega C_1}$$

and

$$V_{in} = \frac{E_s (R_{in} + 1/j\omega C_1)}{R_s + R_{in} + 1/j\omega C_1}.$$

Since $V_{out} = g_m V_{be} R_L$, the voltage gain $A_{v(lf)}$ at low frequencies is

$$A_{v(lf)} = V_{out}/V_{in} = \frac{g_m R_L R_{in}}{R_{in} + 1/j\omega C_1} = \frac{A_{v(mf)}}{1 + 1/j\omega C_1 R_{in}}. \qquad (2.16)$$

When the low-frequency gain has fallen by 3 dB on its mid-frequency value,

$$\left| \frac{A_{v(lf)}}{A_{v(mf)}} \right| = \frac{1}{\sqrt{2}} = \frac{1}{\sqrt{[1 + 1/\omega_1^2 C_1^2 R_{in}^2]}} \quad \text{and} \quad \omega_1 = 1/C_1 R_{in}. \qquad (2.17)$$

The expression for the low-frequency voltage gain can be written as

$$A_{v(lf)} = \frac{A_{v(mf)}}{1 - jf_1/f} \qquad (2.18)$$

where f_1 is the lower 3 dB frequency of the amplifier.

If the voltage gain is taken as V_{out}/E_s then

$$A_{v(lf)} = \frac{g_m R_{in} R_L}{R_s + R_{in} + 1/j\omega C_1} = \frac{A_{v(mf)}}{1 + 1/j\omega C_1 (R_s + R_{in})} \qquad (2.19)$$

and the lower 3 dB frequency is

$$f_1 = 1/2\pi C_1 (R_s + R_{in}).$$ (2.20)

Example 2.7

An amplifier has an input resistance of 1 kΩ and is coupled by a capacitor to a 4.7 kΩ resistive source. Calculate the required value of the coupling capacitor to make the lower 3 dB frequency of the circuit equal to 50 Hz. Find also the loss introduced by the coupling capacitor at 25 Hz.

Solution
From equation (2.20)

$$C_1 = \frac{1}{2\pi \times 50 \times 5700} = 0.558 \ \mu F. \quad (Ans.)$$

With this value of coupling capacitor, the load voltage at 25 Hz relative to the mid-frequency value is

$$\left| \frac{V_{l(mf)}}{V_{l(lf)}} \right| = \frac{1}{\sqrt{\left[1 + \dfrac{1}{(2\pi \times 25 \times 0.558 \times 10^{-6} \times 5700)^2} \right]}}$$

$$= \frac{1}{\sqrt{\left(1 + \dfrac{1}{0.25} \right)}} = 0.447 = -7 \ dB. \quad (Ans.)$$

Decoupling of an emitter resistor becomes increasingly ineffective as the frequency is reduced. The overall low-frequency response of a stage will depend upon the time constants of both the coupling and the decoupling circuits. Many integrated-circuit amplifiers do not employ decoupling capacitors because of the large values needed, and their low-frequency response is determined solely by any input and/or output coupling capacitors fitted.

Figure 2.15 shows an amplifier having both input and output coupling capacitors; the overall low-frequency response of the

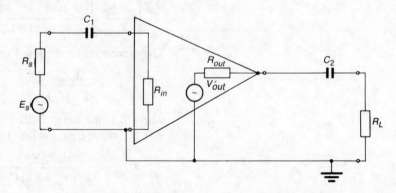

Fig. 2.15 Amplifier with input and output time constants.

amplifier will depend upon the time constants of both the input and the output circuits. Thus

$$A_{v(lf)} = \frac{A_{v(mf)}}{\left(1 + \dfrac{1}{j\omega C_1(R_{in}+R_s)}\right)\left(1 + \dfrac{1}{j\omega C_2(R_{out}+R_L)}\right)}. \quad (2.21)$$

A convenient way of estimating the overall low-frequency gain of an amplifier with more than one time constant is the use of a *Bode plot*.

Bode Plots

The Bode plot of an amplifier consists of two graphs in which the amplitude *in dB* and the phase of the gain are plotted separately to a base of frequency *plotted to a logarithmic scale*. Consider first of all the Bode plot of a single-stage amplifier whose gain–frequency characteristic is given by equation (2.18). The magnitude in dB of $A_{v(lf)}$ is

$$20 \log_{10}|A_{v(lf)}| = 20 \log_{10}|A_{v(mf)}| - 20 \log_{10}\sqrt{[1 + f_1^2/f^2]}.$$

For frequencies above the lower 3 dB frequency f_1, take $|A_{v(lf)}|$ dB as equal to $|A_{v(mf)}|$ dB.

For frequencies below the lower 3 dB frequency where $f < f_1$, the gain is approximately given by

$$20 \log_{10}|A_{v(mf)}| - 20 \log_{10}f_1/f$$

and hence falls with decreasing frequency at a rate of 6 dB per octave (an octave is a doubling of frequency) or 20 dB per decade.

A Bode amplitude plot is drawn in Fig. 2.16(a).

The two lines intersect at $f = f_1$ at which point there is obviously +3 dB error in the plot. Similarly, there is +1 dB error at $f/f_1 = 2$, +1 dB error at $f/f_1 = 0.5$ and negligible error at lower frequencies.

The phase shift θ of the low-frequency gain is

$$\theta = 0° - \tan^{-1} - f_1/f \quad [\text{or } 180° - \tan^{-1} - f_1/f].$$

For frequencies much above f_1, $\theta \simeq 0°$ and when $f = f_1$, $\theta = +45°$. Also, for frequencies less than about $0.1f_1$, θ tends towards $+90°$.

The same principles apply when more than one time constant is involved. For an amplifier with two identical low-frequency time constants, the plot decreases at the rate of 12 dB/octave and the phase plot changes at 90° per decade. If the two time constants differ from one another with, say 3 dB frequencies f_a and f_b with $f_a > f_b$, the Bode amplitude plot is obtained using

$$|A_{v(lf)}| \, \text{dB} = 20\log_{10}|A_{v(mf)}| - 20\log_{10}\sqrt{[1 + (f_a/f)^2]}$$

$$- 20\log_{10}\sqrt{[1 + (f_b/f)^2]}$$

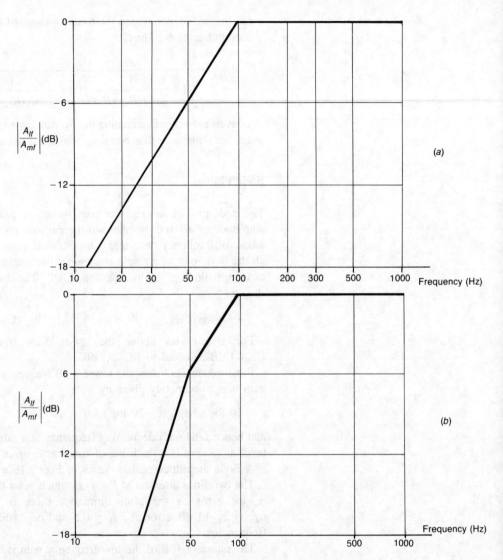

Fig. 2.16 Bode diagrams of amplifier with (*a*) single time constant (*b*) two time constants.

and this means the plot will consist of the sum of:

(*a*) a horizontal straight line at frequencies above f_a;

(*b*) a straight line decreasing at 6 dB/octave starting from 0 dB and f_a; and

(*c*) a straight line with a slope of −6 dB/octave starting from 0 dB and f_b.

This means that at frequencies below f_b the plot has a slope of −12 dB/octave (see Fig. 2.16(*b*)) which supposes $f_a = 100$ Hz and $f_b = 50$ Hz).

To design an amplifier with two low-frequency time constants to have a specified overall lower 3 dB frequency, either of two different methods can be employed.

(*a*) Let one of the time constants determine the required 3 dB point and make the other at least 10 times lower in frequency. This will ensure that the design requirement is achieved before the second time constant starts to have any effect.

(*b*) Let each time constant have equal effect on the low-frequency gain. This means that they should each introduce 1.5 dB loss at the design lower 3 dB point. 1.5 dB is a voltage ratio of 1/1.189 and hence $1.189 = \sqrt{(1+f_1^2/f^2)}$; squaring, $1.413 = 1+f_1^2/f^2$ and $f = 1.55 f_1$. This means that each coupling circuit should have its individual 3 dB point at a frequency that is 1/1.55 times the required overall 3 dB frequency.

Example 2.8

An amplifier has an input resistance of 2000 Ω and an output resistance of 4000 Ω. It is connected between a source of impedance 1000 Ω and a load of 4000 Ω. Determine the necessary values of input and output coupling capacitors for the lower 3 dB frequency to be 25 Hz.

Solution
Method (a)
Set the input circuit to have 3 dB loss at 25 Hz and the output circuit to have its 3 dB point at 2.5 Hz. Then, from equation (2.20),

$$C_1 = \frac{1}{2\pi \times 25 \times 3000} \simeq 2 \ \mu\text{F} \quad (Ans.)$$

$$C_2 = \frac{1}{2\pi \times 2.5 \times 8000} \simeq 8 \ \mu\text{F}. \quad (Ans.)$$

Method (b)
Each circuit should have a lower 3 dB frequency of 25/1.55 = 16.12 Hz. Therefore,

$$C_1 = \frac{1}{2\pi \times 16.12 \times 3000} \simeq 3 \ \mu\text{F} \quad (Ans.)$$

$$C_2 = \frac{1}{2\pi \times 16.12 \times 8000} \simeq 1.2 \ \mu\text{F}. \quad (Ans.)$$

Gain at High Frequencies

The voltage gain of an amplifier will fall at high frequencies because of the unavoidable transistor and stray capacitances. These are all in parallel with one another and so produce effective capacitances in parallel with the input and output signal paths. The total output capacitance will appear in parallel with the input capacitance of the next stage and it is usually considered to be a part of it. Thus the capacitances which affect the high-frequency gain of an amplifier are represented in Fig. 2.17 by C_1 and C_2.

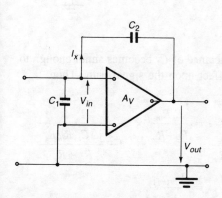

Fig. 2.17 Showing the capacitances that affect the h.f. gain of an amplifier.

The current that flows in C_2 is

$$I_x = (V_{in} - V_{out})j\omega C_2 = (V_{in} + A_v V_{in})j\omega C_2,$$

and hence

$$I_x/V_{in} = (1 + A_v)j\omega C_2.$$

The input admittance Y_{in} of the circuit is

$$Y_{in} = j\omega C_1 + (1 + A_v)j\omega C_2$$

and so the effective input capacitance C_{in} of the amplifier is

$$C_{in} = C_1 + (1 + A_v)C_2 \simeq C_1 + A_v C_2. \tag{2.22}$$

The increase in the input capacitance of the amplifier can be quite large and it is known as the *Miller effect*.

The effective input capacitance of an amplifier will make its voltage gain fall at the higher frequencies. Figure 2.18 shows an amplifier of input resistance R_{in}, effective input capacitance C_{in}, and voltage gain A_v connected to a source of resistance R_s and e.m.f. E_s. At low and middle frequencies where the reactance of the input capacitance is very high the input voltage to the amplifier is $V_{in(mf)} = E_s R_{in}/(R_s + R_{in})$.

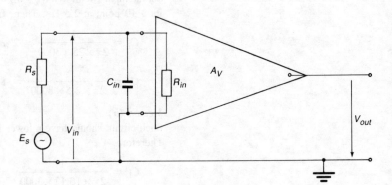

Fig. 2.18 Input capacitance of an amplifier.

At high frequencies the reactance of C_1 becomes small enough to have a noticeable shunting effect upon the signal path. Then,

$$V_{in(hf)} = \frac{E_s \dfrac{R_{in}}{1 + j\omega C_{in} R_{in}}}{R_s + \dfrac{R_{in}}{1 + j\omega C_{in} R_{in}}} = \frac{E_s R_{in}}{R_s + R_{in}} \times \frac{1}{1 + \dfrac{j\omega C_{in} R_{in} R_s}{R_{in} + R_s}}$$

$$= \frac{V_{in(mf)}}{1 + j\omega\tau}.$$

The overall voltage gain $A_{v(hf)} = V_{out}/E_s$ at high frequencies can be written as

$$A_{v(hf)} = \frac{A_{v(mf)}}{1+j\omega\tau} \tag{2.23}$$

$$= \frac{A_{v(mf)}}{1+jf/f_2} \tag{2.24}$$

where f_2 is the upper 3 dB frequency.

Example 2.9

An inverting amplifier has an input resistance of 1500 Ω, an input capacitance of 20 pF, an input–output capacitance of 2 pF, and a voltage gain of 100.

A source of internal resistance 1000 Ω is connected to the amplifier. Calculate the frequencies at which the output voltage is (a) 3 dB, (b) 6 dB, and (c) 10 dB down on its mid-band value.

Solution

From equation (2.22), $C_{in} = 20 + 100 \times 2 = 220$ pF. The input time constant is $200 \times 10^{-12} \times (1000 \times 1500/2500) = 132$ ns. From equation (2.23),

(a) 3 dB fall:

$$\frac{1}{\sqrt{2}} = \frac{1}{\sqrt{(1+\omega_{3\,dB}^2\,\tau^2)}} \quad \text{or} \quad \omega_{3\,dB}\tau = 1$$

and

$$f_{3\,dB} = \frac{1}{2\pi \times 132 \times 10^{-9}} = 1.206 \text{ MHz.} \quad (Ans.)$$

(b) 6 dB fall:

$$\frac{1}{2} = \frac{1}{\sqrt{(1+\omega_{6\,dB}^2\tau^2)}} \quad \text{or} \quad \omega_{6\,dB}\tau = \sqrt{3}$$

and

$$f_{6\,dB} = \sqrt{3}f_{3\,dB} = 2.089 \text{ MHz.} \quad (Ans.)$$

(c) 10 dB fall:

$$\frac{1}{\sqrt{10}} = \frac{1}{\sqrt{(1+\omega_{10\,dB}^2\tau^2)}} \quad \text{or} \quad \omega_{10\,dB}\tau = 3$$

and

$$f_{10\,dB} = 3f_{3\,dB} = 3.618 \text{ MHz.} \quad (Ans.)$$

The load R_L will be shunted by a capacitance C equal to the sum of the transistor output capacitance and the stray capacitances, see Fig. 2.19. The output time constant $\tau = CR_{out}R_L/(R_{out}+R_L)$ will also reduce the gain of the circuit at higher frequencies, producing a 3 dB fall in gain at a frequency $f_{3\,dB} = 1/(2\pi \times \tau)$. The high-frequency response of an amplifier can also be shown by a Bode plot. From equation (2.24) the magnitude in dB of $A_{v(hf)}$ is

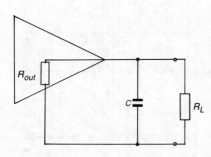

Fig. 2.19 Output time constant of an amplifier.

$20 \log_{10}|A_{v(hf)}| = 20 \log_{10}|A_{v(mf)}| - 20 \log_{10}\sqrt{(1+f^2/f_2^2)}$. For frequencies up to the upper 3 dB frequency f_2, take $|A_{v(hf)}|$ dB as being equal to $|A_{v(mf)}|$ dB. For frequencies above the upper 3 dB frequency where $f > f_2$ the gain is given approximately by

$$20 \log_{10}|A_{v(mf)}| - 20 \log_{10}f/f_2$$

and hence falls at a rate of 6 dB/octave with increasing frequency, (or at 20 dB/decade). The Bode amplitude plot is given by Fig. 2.20(a). The two lines intersect at $f = f_2$ at which point there is +3 dB error in the plot. Also there is +1 dB error at $f/f_2 = 0.5$, +1 dB error at $f/f_2 = 2$ and negligible error at all higher frequencies. The phase shift of the high-frequency gain is given by

$$\theta = 0° - \tan^{-1}f/f_2 \quad [\text{or } 180° - \tan^{-1}f/f_2].$$

For frequencies very much below f_2, $\theta \simeq 0°$ and when $f = f_2$, $\theta = -45°$. Also, for frequencies greater than about $10f_2$, θ tends towards $-90°$. The Bode phase plot is drawn with a slope of 45° per decade in Fig. 2.20(b). It can be determined that this idealized phase characteristic is never in error by more than 6°.

The same principles apply when more than one time constant is involved. For an amplifier with two identical high-frequency time constants, the line falls at the rate of 12 dB/octave (40 dB/decade) and the phase plot changes at 90°/decade. If the two time constants differ from one another, the Bode amplitude plot is obtained using

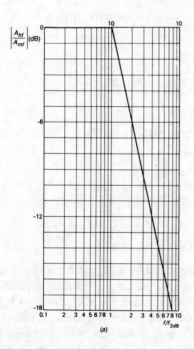

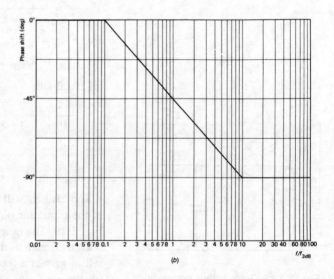

Fig. 2.20 Bode amplitude and phase diagrams.

$$|A_{v(hf)}| \text{ dB} = 20 \log_{10}|A_{v(mf)}| - 20 \log_{10}\sqrt{(1+f^2/f_a^2)}$$
$$- 20 \log_{10}\sqrt{(1+f^2/f_b^2)}$$

and this means that the plot will consist of the sum of

(a) a horizontal line up to $f = f_a$,
(b) a straight line decreasing at 20 dB/decade starting from f_a, and
(c) a straight line with a slope of -20 dB/decade starting from f_b.

The three individual lines are shown dotted, and their resultant solid, in Fig. 2.21(a).

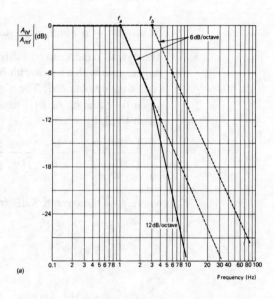

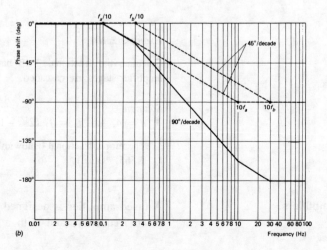

Fig. 2.21 Bode plots for an amplifier with two high-frequency time constants.

Similarly, the phase response can be obtained;

$$\theta = 0° - \tan^{-1}f/f_a - \tan^{-1}f/f_b$$

The phase plot is represented by the sum of

(a) a horizontal line $\theta = 0°$ from $f = 0$ to $f = f_a/10$,
(b) a straight line with $-45°$/decade slope from $f = f_a/10$ to $f = 10f_a$,
(c) a third straight line also with $-45°$/decade slope from $f_b/10$ to $10f_b$, and
(d) horizontal straight lines from $10f_a$, and $10f_b$, upwards in frequency.

These four lines are drawn dotted in Fig. 2.21(b) and their resultant is shown by the solid line.

Bandwidth of a Multi-stage Amplifier

The overall bandwidth of a direct-coupled amplifier with n identical stages, each with a bandwidth B determined by a single time constant can be easily obtained. The high-frequency gain of a single stage is given by equation (2.24), hence for n stages the overall gain has fallen by 3 dB when

$$\left| \frac{A_{v(hf)}}{A_{v(mf)}} \right| = \frac{1}{\sqrt{2}} = \frac{1}{[1+f_0^2/f_2^2]^{n/2}}$$

where f_0 is the overall 3 dB frequency.

$$2 = [1+f_0^2/f_2^2]^n$$
$$2^{1/n} = 1+f_0^2/f_2^2$$

or

$$f_0 = f_2\sqrt{(2^{1/n}-1)}. \tag{2.25}$$

Example 2.10

Calculate the fractional 3 dB bandwidth reduction when four identical amplifier stages are cascaded.

Solution

$n = 4$, so $f_0 = f_2\sqrt{(2^{1/4}-1)} = 0.435$.

Therefore, fractional bandwidth reduction = $(f_2-f_0)/f_2 = 1-0.435 = 0.565$. (*Ans.*)

Tuned Amplifiers

A tuned amplifier is designed to give a specified power gain at a given frequency, and with a specified bandwidth. The required selectivity is usually obtained by the use of a parallel-tuned *LC* circuit as the collector load instead of a resistor. At the higher frequencies one of the main problems is to ensure that the circuit is not unstable and prone to unwanted oscillations. This problem can be overcome by making the transistor work into a load that is somewhat less than the value needed for maximum power transfer. In most cases a bipolar transistor operated in its common-emitter connection is employed, but sometimes either a common-base transistor or a FET is used instead. The FET has the advantage of being less affected by cross-modulation and intermodulation and having the same noise performance; the common-base circuit allows a given type of transistor to be used at a higher frequency than would be possible in common-emitter. Also, a number of IC r.f. amplifiers are available although very often the package also contains one, or more, other circuit functions.

At the lower radio frequencies the tuned circuit can merely replace the collector resistor in an audio circuit and the main differences occur in the way in which a transistor is coupled to the next stage or to the load. The tuned circuit must have a sufficiently high value of *Q* factor to obtain the wanted 3 dB bandwidth and often this will be reduced to too low a value if one stage is capacitor coupled to the next. Some of the coupling methods employed are shown in Fig. 2.22. In each case either the tapping ratio or the turns ratio is chosen to step up the input resistance of the following stage to the value that will give the wanted effective *Q* factor.

At the resonant frequency, $f_0 = 1/2\pi\sqrt{LC}$ Hz, the impedance of the tuned circuit is equal to the dynamic resistance R_d and it is then at its maximum possible value of $L/CR = Q\omega_0 L = Q/\omega_0 C$ ohms.

At any frequency off resonance the tuned circuit's impedance falls and is given by

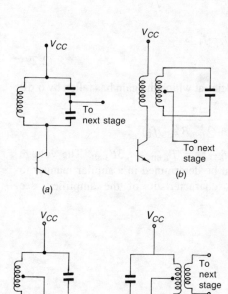

Fig. 2.22 Coupling methods for r.f. amplifiers.

$$Z_L = \frac{R_d}{1 + jQB/f_0}.\tag{2.26}$$

In this equation, Q is the Q factor of the tuned circuit, f_0 is the resonant frequency, and B is the bandwidth that is considered.

The impedance of the tuned circuit is shunted by the resistance referred from the next stage, $n^2 R_{in}$, where n is the tapping or the turns ratio and R_{in} is the input resistance of the next stage. The effective dynamic resistance $R_{d(eff)}$ is hence equal to

$$\frac{R_d n^2 R_{in}}{R_d + n^2 R_{in}}$$

and the effective Q factor $Q_{(eff)}$ is

$$Q_{(eff)} = \frac{R_{d(eff)}}{\omega_0 L} = \frac{Q}{1 + R_d/n^2 R_{in}}.\tag{2.27}$$

The voltage gain of the amplifier is then

$$A_v = g_m Z_L = \frac{g_m R_{d(eff)}}{1 + j Q_{(eff)} B/f_0}. \tag{2.28}$$

At the two frequencies f_1 and f_2 at which the gain has fallen by 3 dB

$$\left| \frac{Z_{L(eff)}}{R_{d(eff)}} \right| = \frac{1}{\sqrt{2}} = \frac{1}{\sqrt{\left(1 + \dfrac{Q_{(eff)}^2 B_{3\,dB}^2}{f_0^2} \right)}}$$

and so the 3 dB bandwidth $f_2 - f_1$ is

$$B_{3\,dB} = f_0 / Q_{(eff)}. \tag{2.29}$$

Similarly, at the two frequencies at which the gain has fallen by 6 dB

$$\left| \frac{Z_{L(eff)}}{R_{d(eff)}} \right| = \frac{1}{2} = \frac{1}{\sqrt{(1 + Q_{(eff)}^2 B_{6\,dB}^2 / f_0^2)}}$$

and the 6 dB bandwidth is $f_b - f_a = B_{6\,dB} = \sqrt{3} B_{3\,dB}$. The voltage gain at other frequencies can be determined in a similar manner to obtain the gain–frequency characteristic of the amplifier, see Fig. 2.23.

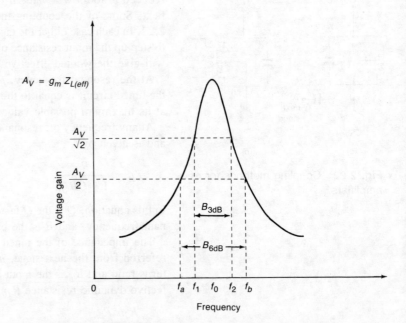

Fig. 2.23 Gain-frequency characteristic for an r.f. amplifier.

Current sources and current mirrors   There are many applications in electronics where a source of constant current is required. The term *constant current* implies that the source will be able to supply the same current to a load regardless (within limits) of the value of that load. In turn, this implies the use of a circuit that has a very high output impedance to signal-frequency currents. The symbol for a current source is given in Fig. 2.24(a).

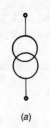

(a)

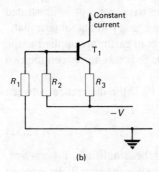

(b)

Fig. 2.24 (a) Symbol for a current source. (b) Basic transistor current source.

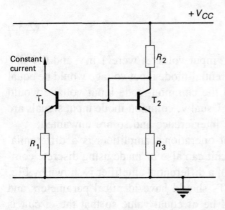

Fig. 2.25 Current source.

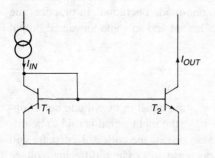

Fig. 2.26 Current mirror.

A variety of circuits can be used as current sources and Fig. 2.24(b) shows one of them. Resistors R_1 and R_2 hold the base of T_1 at a constant voltage V_B. The voltage across the emitter resistor R_3 is equal to $V_B - V_{BE}$ and so the current that flows in R_3 is constant at $(V_B - V_{BE})/R_3$. The collector current will therefore be held constant at very nearly this value regardless of the resistance in which it flows. In practice, two diodes are connected in series with R_2 to counteract the temperature variations of V_{BE}.

Another circuit is given in Fig. 2.25. The collector current passed by T_2 is

$$I_{C2} = \frac{V_{CC} - V_{BE2}}{R_2 + R_3}$$

and so the voltage developed across R_3 is

$$(V_{CC} - V_{BE2})R_3/(R_2 + R_3).$$

Hence,

$$\frac{(V_{CC} - V_{BE2})R_3}{R_3 + R_2} + V_{BE2} = V_{BE1} + (I_{B1} + I_{C1})R_1.$$

Now, $I_{C1} \gg I_{B1}$ and $V_{BE2} \simeq V_{BE1}$ and so

$$I_{C1} = \frac{(V_{CC} - V_{BE2})R_3}{R_1(R_2 + R_3)}. \tag{2.30}$$

Transistor T_1 has negative feedback applied to it and this increases its output impedance to a high value.

Current Mirror

The basic circuit of a current mirror is shown by Fig. 2.26. It consists of two transistors, one of which is connected as a diode, that have their base and emitter terminals connected together. The two transistors must be matched devices and are usually within an integrated circuit. For T_2, $I_B + I_C = I_E$, so that

$$I_C = I_{OUT} = I_E - I_B = I_E - \frac{I_E}{1 + h_{FE}} = I_E[1 - 1/(1 + h_{FE})]$$

$$= h_{FE}I_E/(1 + h_{FE}).$$

For T_1,

$$I_{IN} = (V_{CC} - V_{BE1})/R_1 = h_{FE}I_E/(1 + h_{FE}) + 2I_E/(1 + h_{FE})$$

$$= (h_{FE} + 2)I_E/(1 + h_{FE}).$$

The current gain of the circuit is $A_i = I_{OUT}/I_{IN} = h_{FE}/(h_{FE} + 2)$. If $h_{FE} \gg 2$ then $I_{OUT} = I_{IN}$ and the action of the circuit has been to reproduce the input current at the output. Since the variation of V_{BE}

with temperature is usually fairly small compared with the supply voltage V_{CC} the repeated current is very nearly independent of temperature (V_{CC}/R_1).

The main application of the current mirror is in conjunction with bias arrangements in integrated circuits.

The Differential Amplifier

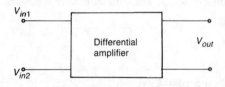

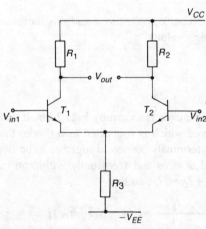

Fig. 2.27 Differential amplifier.

The function of a differential amplifier is to amplify the *difference* between two input signal voltages. The basic concept is illustrated by Fig. 2.27. The amplifier has two input, and two output terminals. Input voltages V_{in1} and V_{in2}, with respect to earth, are applied to the two input terminals and the output voltage is taken between the two output terminals.

The *differential-mode input voltage* V_d is the difference between the two input voltages, i.e.

$$V_d = V_{in1} - V_{in2} \tag{2.31}$$

and ideally, only this voltage should be amplified. In practice, however, a further component of output voltage will be present which is proportional to the average of the two input signal voltages. This average signal voltage is known as the *common-mode input voltage* V_c, thus

$$V_c = \frac{V_{in1} + V_{in2}}{2}. \tag{2.32}$$

If, for example, the two input voltages were 1 mV and 0.9 mV respectively, then the differential-mode input voltage would be equal to $1 - 0.9 = 0.1$ mV, and the common-mode input voltage would be $(1+0.9)/2 = 0.95$ mV. Usually, common-mode input signals are generated by noise and/or interference and so are unwanted.

The input stage of most operational amplifiers is a differential amplifier stage and the circuit can also be made using discrete components. The basic circuit of a differential amplifier is shown by Fig. 2.28. Transistors T_1 and T_2 should have identical parameters and resistors R_1 and R_2 should be of equal value so that the circuit is balanced. If the two input voltages are equal in both magnitude and polarity, equal collector currents will flow and the output voltage will be zero. This is the common-mode operation. In practice, the circuit will *not* be perfectly balanced and so some unwanted output voltage will exist.

Fig. 2.28 Basic differential amplifier circuit.

Differential Input

If the input voltages V_{in1} and V_{in2} are equal in magnitude but of opposite polarity, i.e. $V_{in1} = -V_{in2}$, the input signal is said to be differential. During each half cycle one of the collector currents will increase and the other will decrease to produce differing voltage drops across the collector resistors R_1 and R_2, and hence will develop an output voltage. The sum of the emitter currents of T_1

and T_2 flows in the common emitter resistor R_3 and is of constant value so that zero signal voltage is developed across R_3. This means that there is no need for an emitter decoupling capacitor. The signal-frequency current flows into the base of T_1, out of its emitter, and into the emitter of T_2. This means that T_2 is effectively operated as a common-base stage with an input resistance of h_{ib} ohms, and T_1 is operated as a common-emitter stage with an undecoupled emitter resistance of h_{ib}. Hence the voltage gain of T_1 is $h_{fe}R_1/(h_{ie}+h_{fe}h_{ib})$ (see eqn 2.7).

Since $h_{ib} = h_{ie}/h_{fe}$,

$$A_v = \frac{h_{fe}R_1}{2h_{ie}} = \frac{g_m R_1}{2}. \tag{2.33}$$

This is the gain from one input terminal to one output terminal. Since $V_{in1} = -V_{in2}$ the differential input voltage is $2V_{in1}$ and so the differential-mode voltage gain A_{vd} is twice as large, i.e.

$$A_{vd} = g_m R_1. \tag{2.34}$$

The input resistance of the circuit to differential-mode signals is

$$R_{ind} = h_{ie}+h_{fe}h_{ib} = 2h_{ie}. \tag{2.35}$$

Common-mode Input

When a common-mode input voltage is applied to the circuit of Fig. 2.28 the two transistors effectively act in parallel with one another. The signal-frequency current flowing in the emitter resistor R_3 is then very nearly equal to V_{in}/R_3 and so the a.c. collector current of each transistor is $V_{in}/2R_3$. These currents produce an a.c. output voltage of $I_c R_1$ at each output terminal, so that the common-mode voltage gain is

$$A_{vc} = \frac{V_{in}R_1/2R_3}{V_{in}} = \frac{R_1}{2R_3}. \tag{2.36}$$

It is desirable that A_{vc} is as small as possible to reduce the unwanted common-mode output voltage and this requires the use of a large value of emitter resistor R_3.

Common-mode Rejection Ratio

The ability of a differential amplifier to reject common-mode signals is expressed by its *common-mode rejection ratio (c.m.r.r.)*.

$$\text{c.m.m.r} = \frac{\text{differential-mode voltage gain } A_{vd}}{\text{common-mode voltage gain } A_{vc}} \tag{2.37}$$

$$= \frac{g_m R_1/2}{R_1/2R_3} = g_m R_3. \tag{2.38}$$

The c.m.m.r. is usually quoted in decibels.

Example 2.11

A differential amplifier has collector resistors of 4700 ohms, an emitter resistor of 12 kΩ, and the transistors have a mutual conductance g_m of 40 mS. Calculate (*a*) its c.m.m.r., (*b*) its output voltage when input signals of 100 mV and 90 mV are applied, and (*c*) its output voltage when input signals of +100 mV and −100 mV are applied.

Solution

(*a*) Common-mode rejection ratio $= 40 \times 10^{-3} \times 12 \times 10^3 = 480$ or 53.6 dB. (*Ans.*)

(*b*) $A_v = \dfrac{40 \times 10^{-3} \times 4700}{2} = 94$

$A_{vc} = 4700/(24 \times 10^3) = 0.196.$

Therefore, the output voltage is

$$94 \times 10 \times 10^{-3} + 0.196 \times 95 \times 10^{-3} = 958.6 \text{ mV.} \quad (Ans.)$$

(*c*) Now the effective differential input voltage is 200 mV and the common-mode input voltage is zero. Hence the output voltage is

$$94 \times 200 \times 10^{-3} = 18.8 \text{ V.} \quad (Ans.)$$

Emitter Resistance

The c.m.m.r. of a differential amplifier is proportional to the value of the emitter resistance. The value of the *resistor* which can be used is limited by the d.c. voltage that will be dropped across it. This difficulty can, to some extent, be overcome by increasing the value of $-V_{EE}$ but it is better to use a constant-current source to provide the emitter resistance.

Variations on the Basic Differential Amplifier

Single-input, Differential Output
One input, say V_{in2}, is now at earth potential. The differential input voltage is $V_d = V_{in1} - V_{in2} = V_{in1}$ and the output is $V_{out} = A_{vd} V_{in1} = (g_m R_1/2) V_{in1}.$

Single-input, Single Output
The voltage gain is obtained by putting $V_{in2} = 0$ in the earlier equations, so that $V_d = V_{in1}$ and the output voltage is $V_{out} = A_{vd} V_{in1} = g_m R_1 V_{in1}/2.$

The output voltage may be taken from either collector; if taken from T_1 a phase reversal occurs, if taken from T_2 there is no phase reversal.

3 Negative Feedback

An audio-frequency amplifier can be designed to have a certain voltage, current, or power gain together with particular values of input and output impedance. The amplifier will add noise and distortion to any signal it amplifies. The components used in the amplifier, both passive and active (transistors and ICs), will vary in value both with time and with variation in temperature and also when a component has to be replaced by another of the same type and nominal value. The parameters of a transistor, such as its mutual conductance, depend upon the operating conditions and so any fluctuations in the power supplies may also cause the gain of the amplifier to alter. For many applications it is important that the gain of an amplifier be kept as constant as possible and then *negative feedback* (n.f.b.) is applied to the amplifier, at the expense of a reduction in the gain.

The general block diagram of a negative feedback amplifier is shown in Fig. 3.1. A feedback network is connected in parallel with the output terminals of the amplifier so that a fraction of the output voltage can be fed back to the input of the amplifier. The total voltage applied to the input of the amplifier proper is the phasor sum of the input voltage and the fed-back voltage.

If the fed-back voltage is in antiphase with the input voltage, the feedback is *negative* and the overall gain of the amplifier will be reduced. Conversely, if the input and fed-back voltages are in phase with one another, *positive feedback* is applied to the amplifier and its gain will be increased. Positive feedback will lead to the circuit becoming

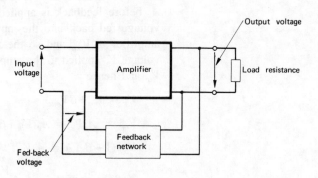

Fig. 3.1 Block diagram of a feedback amplifier.

unstable and prone to oscillate at some particular frequency, and it is the basis of an oscillator circuit.

At middle frequencies where the reactances of the circuit capacitances can be neglected, it is easy to ensure that the feedback is negative. At low and high frequencies however, phase shifts within the amplifier caused by capacitive reactances will ensure that the fed-back voltage is no longer in antiphase with the input voltage. It may often not be immediately evident whether the feedback is now negative or positive. If the internal phase shifts are large enough, negative feedback may be turned into positive feedback at certain low and high frequencies and there is then a danger that the circuit may oscillate.

Differences exist in the methods used to derive the fed-back signal and to introduce it into the input circuit. These differences lead to the classification of n.f.b. amplifiers into one of four types: (*a*) voltage-series; (*b*) voltage-shunt; (*c*) current-series; and (*d*) current-shunt.

In general, these various types of n.f.b. have similar effects upon the performance of an amplifier other than on its input and output impedances. Each impedance may either be increased or decreased according to the type of n.f.b. (see Table 3.1). Note that the changes in input and output impedance are not in the same ratio.

Table 3.1 Types of negative feedback and effect on impedances

Type of n.f.b.	Input Impedance	Output Impedance
Voltage-series	Increased	Decreased
Voltage-shunt	Increased	Increased
Current-shunt	Decreased	Increased
Current-series	Decreased	Decreased

Voltage-series Feedback

With voltage-series feedback, a voltage that is proportional to the output voltage of the amplifier, is fed back and applied in series with the input signal voltage (Fig. 3.2). The amplifier has a voltage gain A_v before feedback is applied and β is the fraction of the output voltage fed back into the input circuit. The output voltage of the amplifier is V_{out} and so the fed-back voltage is βV_{out}. The total voltage V_x applied to the amplifier is the phasor sum of βV_{out} and V_{in} and hence

$$V_x = V_{in} + \beta V_{out}$$

$$V_{out} = A_v V_x = A_v(V_{in} + \beta V_{out})$$

$$V_{out}(1 - \beta A_v) = A_v V_{in}.$$

$$\text{Voltage gain } A_{v(f)} = \frac{V_{out}}{V_{in}} = \frac{A_v}{1 - \beta A_v}. \tag{3.1}$$

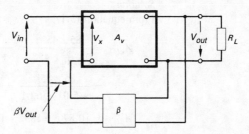

Fig. 3.2 Voltage-series feedback.

If the feedback is *definitely* negative, then *either* β or A_v must be negative so that the loop gain βA is negative and then

$$A_{v(f)} = \frac{A_v}{1-(-\beta A_v)} = A_v/(1+\beta A_v). \qquad (3.2)$$

The requirement for the loop gain to be negative means that, if the amplifier has an even number of phase-inverting stages, its overall gain is $A_v \angle 0°$ and the β network must introduce the required 180° phase shift. If the amplifier has an odd number of phase-inverting stages so that its gain is $A_v \angle 180°$, then zero phase shift must be introduced by the feedback network. Very often the amount of feedback applied to a circuit is quoted in terms of the ratio (gain without n.f.b. to gain with n.f.b.) expressed in dB, i.e. feedback applied is

$$20 \log_{10}|A_v/A_{v(f)}| = 20 \log_{10}|1+\beta A|. \qquad (3.3)$$

Example 3.1

A voltage amplifier has a gain of 120 before n.f.b. is applied. Calculate its voltage gain if 1/10 of the output voltage is fed back to the input in antiphase with the input signal.

Solution
From equation (3.2)

$$A_{v(f)} = 120/(1+120/10) = 923. \quad (Ans.)$$

This example makes it clear that the application of n.f.b. to an amplifier results in a considerable reduction in its gain. At low frequencies the presence of coupling and/or decoupling capacitors will reduce the voltage gain of the amplifier and introduce a phase shift. The loop gain βA of the circuit does not then have a phase angle of 180° and equation (3.1) must be used in any calculation of voltage gain.

Example 3.2

A voltage amplifier has a gain of $85 \angle 45°$ before n.f.b. is applied. Calculate the voltage gain if 1/10 of the output voltage is fed back to the input in antiphase with the input signal.

Solution

From equation (3.1),

$$A_{v(f)} = \frac{85 \angle 45°}{1 - \frac{85}{10} \angle 45°} = \frac{85 \angle 45°}{1 - 8.5\cos45° - j8.5\sin45°}$$

$$= \frac{85 \angle 45°}{-5.01 - j6.01} = \frac{85 \angle 45°}{7.82 \angle 230°}$$

$$A_{v(f)} = 10.87 \angle -185°. \quad (Ans.)$$

The overall voltage gain of an amplifier is influenced by the relative values of the source and amplifier input impedances. The overall voltage gain $A_{vo(f)}$ is the ratio V_{out}/E_s.

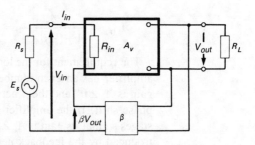

Fig. 3.3

Referring to Fig. 3.3,

$$E_s = I_{in}(R_s + R_{in}) - \beta V_{out}$$

$$= \frac{V_{in}}{R_{in}}(R_s + R_{in}) - \beta V_{out}.$$

$$A_{vo(f)} = V_{out}/E_s$$

$$= \frac{V_{out}}{\dfrac{V_{in}}{R_{in}}(R_s + R_{in}) - \beta V_{out}} = \frac{V_{out}}{V_{in}\left(\dfrac{R_s + R_{in}}{R_{in}} - \beta A_v\right)}$$

$$= \frac{A_v}{\dfrac{R_s + R_{in}}{R_{in}} - \beta A_v}. \tag{3.4}$$

Clearly if $R_s \ll R_{in}$, this expression reduces to equation (3.1).

Input Impedance

In the derivation of the expressions for the input and output impedances of a feedback amplifier, it will be assumed that the feedback is negative and that the impedances are resistive.

The input impedance of an amplifier is the ratio V_{in}/I_{in}. From Fig. 3.4,

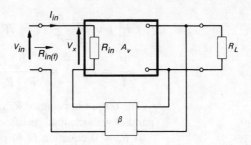

Fig. 3.4

$$I_{in} = \frac{V_{in}}{R_{in(f)}} = \frac{V_x}{R_{in}}$$

but $V_x = V_{in} - \beta A_v V_x$ or $V_{in} = V_x(1+\beta A_v)$.

Hence

$$R_{in(f)} = \frac{V_{in}}{I_{in}} = \frac{V_x}{I_{in}}(1+\beta A_v)$$

$$R_{in(f)} = R_{in}(1+\beta A_v). \tag{3.5}$$

Equation (3.5) shows that the input impedance of a voltage-series feedback amplifier is *increased*.

Output Impedance

The output resistance of an n.f.b. amplifier is best determined by replacing the input signal source by an impedance equal to its internal resistance R_s, and connecting a voltage generator of e.m.f. E and zero internal impedance across the output terminals of the amplifier (see Fig. 3.5). The output impedance $R_{out(f)}^1$ of the amplifier with the load disconnected is then given by the ratio E/I.

R_{out} is the output impedance of the amplifier before negative feedback is applied and A_v^1 is the *open-circuit* voltage gain of the amplifier, i.e. the voltage gain with $R_L = \infty$.

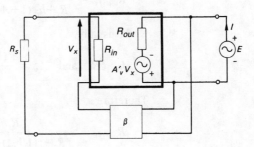

Fig. 3.5

From Fig. 3.5

$$E+A_v^1 V_x = IR_{out}$$

but $V_x = \beta E R_{in}/(R_s+R_{in})$ so that

$$E\left[1+\frac{\beta A_v^1 R_{in}}{R_s+R_{in}}\right] = IR_{out}$$

$$\frac{E}{I} = R_{out(f)}^1 = \frac{R_{out}(R_s+R_{in})}{R_s+R_{in}(1+\beta A_v^1)}. \tag{3.6}$$

The output impedance $R_{out(f)}$ of the amplifier is equal to $R_{out(f)}^1$ in parallel with the load resistance. Note that A_v^1 is V_{out}/V_x, *not* V_{out}/E_s.
If $R_s = 0$ equation (3.6) reduces to

$$R_{out(f)}^1 = \frac{R_{out}}{1+\beta A_v^1}. \tag{3.7}$$

Equation (3.7) can also be used to determine $R_{out(f)}$ when $R_s \neq 0$ if A_v^1 is defined as the open-circuit gain V_{out}/E_s instead of V_{out}/V_x (see Example 3.3).

An example of voltage-series n.f.b. is the emitter follower (page 49). The output voltage is in series with the input signal voltage but with the opposite polarity, thus *all* of the output voltage is applied as negative feedback ($\beta = 1$). From equation (3.2),

$$A_{v(f)} = \frac{A_i R_{L(eff)}/h_{ie}}{1+A_i R_{L(eff)}/h_{ie}} = \frac{A_i R_{L(eff)}}{h_{ie}+A_i R_{L(eff)}}. \tag{3.8}$$

Equation (3.8) should be compared with equation (2.11).
From equation (3.5)

$$R_{in(eff)} = h_{ie}\left(1+\frac{A_i R_{L(eff)}}{h_{ie}}\right).$$

$$= h_{ie}+A_i R_{L(eff)} \simeq A_i R_{L(eff)}. \tag{3.9}$$

[See eqn (2.10).]

Finally, the output resistance of an emitter follower is (eqn 3.6),

$$R_{out(f)} = \frac{(1/h_{oe})(h_{ie}+R_s)}{R_s+h_{ie}(1+A_v^1)} \simeq \frac{(1/h_{oe})(h_{ie}+R_s)}{R_s+A_v^1 h_{ie}}.$$

Now, usually,

$$h_{ie} \ll R_s \quad \text{and} \quad A_v^1 = \frac{(1/h_{oe})+R_{L(eff)}}{R_{L(eff)}} \times \frac{h_{fe}R_{L(eff)}}{h_{ie}}.$$

Therefore

$$R_{out(f)} = \frac{R_s/h_{oe}}{R_s+[(1/h_{oe})+R_{L(eff)}]h_{fe}}$$

$$\simeq \frac{R_s/h_{oe}}{R_s+h_{fe}/h_{oe}} \simeq \frac{R_s/h_{oe}}{h_{fe}/h_{oe}} = R_s/h_{fe}. \tag{3.10}$$

Example 3.3

An amplifier has a voltage gain of 112, an input resistance of 2 kΩ, an output resistance of 6 kΩ and a load resistance of 2000 Ω. The amplifier has negative feedback with $\beta = 0.01$ applied and is used to amplify the signal provided by a source of e.m.f. 10 mV and impedance 1500 Ω. Calculate

(a) the output resistance of the amplifier, and
(b) the voltage developed across the load resistance.

Solution
Fig. 3.6 shows the circuit.

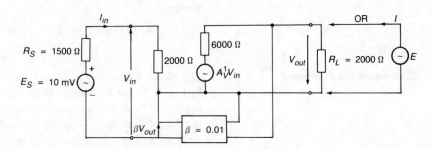

Fig. 3.6

Method 1
(a) With the load resistance R_L removed and replaced by the generator of E volts e.m.f., and the source e.m.f. set to zero,

$$R_{out(f)} = E/I \quad \text{and} \quad V_{in} = \beta E R_{in}/(R_s+R_{in}) = 0.01E \times 2000/3500$$

$$= 5.714 \times 10^{-3} E$$

$A_v^1 = 112 \times 8000/2000 = 448$, so

$$I = \frac{E+448 \times 5.714 \times 10^{-3} E}{6000} = \frac{3.56E}{6000}.$$

Therefore,

$$R_{out(f)} = \frac{E}{I} = \frac{6000}{3.56} = 1685 \ \Omega. \quad (Ans.)$$

(b) $\quad V_{in} = \dfrac{(E_s - \beta V_{out})R_{in}}{R_s + R_{in}} \quad V_{out} = A_v V_{in}$

$$V_{out}\left(1 + \frac{\beta R_{in} A_v}{R_s + R_{in}}\right) = \frac{A_v E_s R_{in}}{R_s + R_{in}}$$

$$V_{out} = \frac{A_v E_s R_{in}}{R_s + R_{in}(1+\beta A_v)} = \frac{112 \times 10 \times 10^{-3} \times 2000}{1500 + 2000(1 + 0.01 \times 112)} = 390 \ \text{mV}.$$

$$(Ans.)$$

Method 2
(a) With R_L disconnected, from equation (3.6)

$$R_{out(f)} = \frac{6000 \times 3500}{1500 + 2000(1 + 0.01 \times 448)} = 1685 \ \Omega. \quad (Ans.)$$

(b) From equation (3.5)

$$R_{in(f)} = R_{in}(1+\beta A_v) = 2000(1+0.01 \times 112) = 4240 \ \Omega.$$

From equation (3.2)

$$A_{v(f)} = \frac{112}{1+0.01 \times 112} = 52.83$$

$$V_{in} = \frac{E_s R_{in(f)}}{R_s + R_{in(f)}} = \frac{10 \times 10^{-3} \times 4240}{1500 + 4240} = 7.387 \ \text{mV}.$$

Therefore,

$$V_{out} = 7.387 \times 10^{-3} \times 52.83 = 390 \ \text{mV}. \quad (Ans.)$$

Current-series Feedback

The block diagram of a current-series feedback amplifier is shown by Fig. 3.7. The fed-back voltage is proportional to the current flowing in the load and is applied in series with the input signal. Since the fed-back voltage is applied in series with the input circuit, the analysis leading to equations (3.2) and (3.5) is again applicable. Hence, the voltage gain $A_v = V_{out}/V_{in}$ is again $A_{v(f)} = A_v/(1+\beta A_v)$.

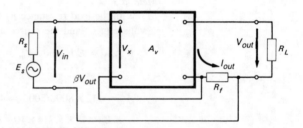

Fig. 3.7 Current-series feedback.

Also the input impedance is $R_{in(f)} = R_{in}(1+\beta A_v)$.

The feedback voltage βV_{out} is developed by the output current flowing in the feedback resistor R_f. Hence,

$$\beta V_{out} = I_{out} R_f = \frac{V_{out} R_f}{R_L} \quad \text{and so} \quad \beta = R_f/R_L.$$

The simplest way of applying current-series negative feedback to a circuit is merely to leave the emitter resistor undecoupled as shown in Fig 2.6 for a bipolar transistor circuit. Very often $h_{ie} \ll h_{fe} R_4$ and then the gain is approximately given by $R_3 R_4$ or $1/\beta$.

Output Impedance

To derive an expression for the output impedance of a current-series n.f.b. amplifier, the input terminals should be closed in an impedance equal to the source impedance, and a generator of e.m.f. E volts and zero impedance should be connected across the output terminals (fig. 3.8). The output impedance is then E/I ohms.

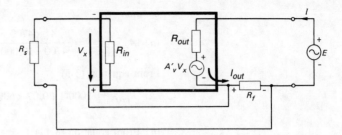

Fig. 3.8

Neglecting R_L for the moment, the voltage equation for the output circuit is

$$E - A_v^1 V_x = I_{out}(R_{out} + R_f).$$

(As before, A_v^1 = open-circuit voltage gain.)

But $V_x = \dfrac{\beta V_{out} R_{in}}{R_s + R_{in}} = \dfrac{I_{out} R_f R_{in}}{R_s + R_{in}}.$

Hence

$$E - \frac{A_v^1 I_{out} R_f R_{in}}{R_s + R_{in}} = I_{out}(R_{out} + R_f)$$

$$E = I_{out}\left(R_{out} + R_f + \frac{A_v^1 R_f R_{in}}{R_s + R_{in}}\right).$$

Therefore,

$$R_{out(f)} = E/I_{out} = R_{out} + R_f\left(1 + \frac{A_v^1 R_{in}}{R_s + R_{in}}\right). \qquad (3.11)$$

If the source resistance can be neglected as small compared with the input impedance R_{in} of the amplifier, this equation reduces to

$$R_{out(f)} = R_{out} + R_f(1 + A_v^1). \qquad (3.12)$$

The feedback can lead to a large increase in the output impedance of the active device but this will be reduced by the shunting effect of R_L.

Example 3.4

An amplifier has an input impedance of 1 kΩ, an output impedance of 10 kΩ and a voltage gain of 600 before the application of n.f.b. Current-series n.f.b. is applied to the amplifier with $\beta = 0.07$.

Calculate (*a*) the input impedance, (*b*) the output impedance, and (*c*) the voltage gains V_{out}/V_{in} and V_{out}/E_s of the amplifier after n.f.b. has been applied. Assume a source resistance of 1000 Ω, a load resistance of 4700 Ω, and a feedback resistor of 330 Ω.

Solution

(a) $A_v = 600 = \dfrac{A_v^1 \times 4700}{4700 + 10\ 000}$ or $A_v^1 = 1877.$

From equation (3.5)

$$R_{in\,(f)} = 1000(1 + 0.07 \times 600) = 43 \text{ k}\Omega. \quad (Ans.)$$

(b) From equation (3.11)

$$R_{out\,(f)} = 10\ 000 + 330\left(1 + \frac{1877 \times 1000}{2000}\right) = 320.04 \text{ k}\Omega. \quad (Ans.)$$

(c) From equation (3.2)

$$A_{v(f)} = \frac{600}{1 + 600 \times 0.07} = 13.95 \quad (Ans.)$$

and from equation (3.4)

$$A_{vo\,(f)} = \frac{600}{\dfrac{1000 + 1000}{1000} + 600 \times 0.07} = 13.64. \quad (Ans.)$$

This last answer can be obtained in an alternative manner. Since the input impedance, with n.f.b. applied, of the amplifier is 43 kΩ, the input voltage V_{in} of the amplifier is

$$V_{in} = E_s \times 43/(43 + 1) = 0.9773 E_s$$

and hence the overall voltage gain is

$$0.9773 \times 13.95 = 13.63. \quad (Ans.)$$

Current-shunt Feedback

With current-shunt feedback, a fraction of the output current is fed back in parallel with the input circuit. The block diagram of the arrangement is shown by Fig. 3.9 in which the directions of the currents shown assume that the feedback is negative. Since a current is fed back it is better to work in terms of the current gain of the amplifier.

From Fig. 3.9

$$\beta I_{out} = \frac{I_{out} R_{f1}}{R_{f1} + R_{f2} + R}$$

where R is the parallel combination of R_s and R_{in}, i.e.

$$R = R_s R_{in}/(R_s + R_{in}).$$

Hence $\beta = R_{f1}/(R_{f1} + R_{f2} + R).$ \hfill (3.13)

Often $R_{f2} \gg R$, then

$$\beta = R_{f1}/(R_{f1} + R_{f2}) \simeq R_{f1}/R_{f2}. \qquad (3.14)$$

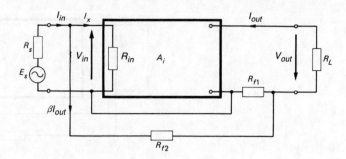

Fig. 3.9 Current-shunt feedback.

Also $I_s = I_{in} - \beta I_{out}$

$$I_{out} = A_i I_s = A_i (I_{in} - \beta I_{out})$$

$$I_{out}(1 + \beta A_i) = A_i I_{in}$$

$$A_{i(f)} = \frac{I_{out}}{I_{in}} = \frac{A_i}{1 + \beta A_i}. \tag{3.15}$$

Input Impedance

The simplest method of deriving an equation for the input impedance of a current-shunt n.f.b. amplifier is to work in terms of admittances. Hence, from Fig. 3.9

$$I_{in} = I_s + \beta I_{out}$$

$$= I_s(1 + \beta A_i)$$

$$\frac{I_{in}}{V_{in}} = Y_{in(f)} = \frac{I_s}{V_{in}}(1 + \beta A_i) = Y_{in}(1 + \beta A_i)$$

$$R_{in(f)} = 1/Y_{in(f)} = R_{in}/(1 + \beta A_i). \tag{3.16}$$

Thus, the input impedance of an amplifier is reduced by the application of current-shunt n.f.b.

Output Impedance

For the calculation of the output impedance with n.f.b., the input terminals of the circuit should be closed with an impedance equal to the source impedance and a voltage generator of e.m.f. E volts and zero internal impedance connected across the output terminals. This is shown by Fig. 3.10 in which A_i^1 is the current gain of the amplifier with its output terminals shorted together. The load resistance R_L has been omitted from the diagram.

The fed-back current is βI where

$$\beta = \frac{R_{f1}}{R_{f1} + R_{f2} + R} \quad \text{with} \quad R = R_s R_{in}/(R_s + R_{in}).$$

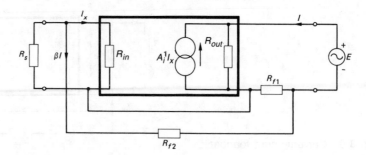

Fig. 3.10

Then $\quad E = R_{out}(I+I_x)+IR_{f1}$

where $\quad I_x = \beta IR_s/(R_s+R_{in})$

$$R_{out(f)} = E/I = R_{out}\left(1+\frac{A_i^1\,R_s}{R_s+R_{in}}\cdot\frac{R_{f1}}{R_{f1}+R_{f2}+R}\right)+R_{f1}\,.$$

$$(3.17)$$

If $R_{f2}\gg R$, $R_{f2}\gg R_{f1}$ and $R_s\gg R_{in}$, this expression reduces to

$$R_{out(f)} = R_{out}\left(1+\frac{A_i^1 R_{f1}}{R_{f2}}\right)+R_{f1}\,.$$

$$(3.18)$$

Thus, the output impedance of the amplifier is increased by the application of current-shunt n.f.b.

Voltage-shunt Feedback

The fourth type of feedback involves feeding back to the input circuit a current whose magnitude is proportional to the output voltage. The block diagram of a voltage-shunt n.f.b. amplifier is shown in Fig. 3.11.

From Fig. 3.11,

$$\beta I_{out} = \frac{V_{out}}{R_f+R}, \quad \text{where} \quad R = R_s R_{in}/(R_s+R_{in}),$$

$$\frac{\beta I_{out}}{I_{out}} = \frac{V_{out}}{(R_f+R)V_{out}/R_L}$$

$$\beta = R_L/(R_f+R).$$

$$(3.19)$$

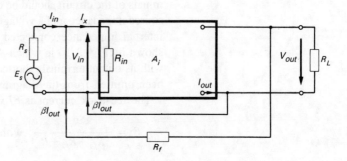

Fig. 3.11 Voltage-shunt feedback.

The current gain and the input impedance with feedback of the amplifier are obtained in the same way as for the current-shunt circuit and leads to the same equations, i.e. (3.15) and (3.16).

Output Impedance

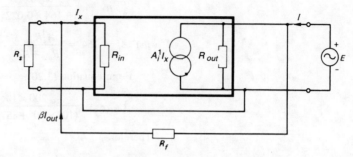

Fig. 3.12

To determine an expression for the output impedance of the amplifier, the circuit given in Fig. 3.12 must be employed. From Fig. 3.12

$$I_x = \frac{E}{R_f + R} \cdot \frac{R_s}{R_s + R_{in}}, \text{ where } R = R_s R_{in}/(R_s + R_{in}),$$

and

$$I_{out} = \frac{A_i^1 E R_s}{(R_f + R)(R_s + R_{in})} + \frac{E}{R_{out}} + \frac{E}{R_f + R}$$

Therefore,

$$Y_{out(f)} = \frac{I_{out}}{E} = \frac{A_i^1 R_s}{(R_f + R)(R_s + R_{in})} + \frac{1}{R_{out}} + \frac{1}{R_f + R}$$

$$\simeq \frac{A_i^1 R_s}{(R_f + R)(R_s + R_{in})} + \frac{1}{R_{out}}. \tag{3.20}$$

Example 3.5

An amplifier has an input resistance of 6 kΩ, a current gain of 164, an output resistance of 10 kΩ and has a load resistance of 2.2 kΩ. The voltage source has an e.m.f. of 10 mV and an internal resistance of 1 kΩ. Voltage-shunt negative feedback is applied to the amplifier using a feedback resistor of 100 kΩ. Calculate (a) the input resistance, (b) the output resistance, (c) the voltage gain of the n.f.b. amplifier, and (d) the signal voltage appearing across the load resistor.

Solution
From equation (3.19)

$$\beta = \frac{2200}{100 \times 10^3 + \frac{6 \times 10^3 \times 10^3}{7 \times 10^3}} = 0.022.$$

(*a*) From equation (3.15)

$$A_{i(f)} = \frac{164}{1 + 164 \times 0.022} = 35.6.$$

From equation (3.16),

$$R_{in(f)} = \frac{6000}{1 + 164 \times 0.022} = 1302 \ \Omega. \quad (Ans.)$$

(*b*) $\ A_i = 164 = \dfrac{A_i^1 R_{out}}{R_{out} + R_L}$, so $A_i^1 = 200$.

From equation (3.20),

$$Y_{out(f)} = \frac{200 \times 1 \times 10^3}{(100 \times 10^3 + 857)(7 \times 10^3)} + 1 \times 10^{-4} = 3.833 \times 10^{-4}.$$

Therefore,

$$R_{out(f)} = \frac{1}{Y_{out(f)}} - 2609 \ \Omega. \quad (Ans.)$$

(*c*) $\ A_{v(f)} = \dfrac{A_{i(f)} R_L}{R_{in(f)}} = \dfrac{35.6 \times 2200}{1302} = 60.15. \quad (Ans.)$

(*d*) $\ V_{in} = \dfrac{E_s R_{in(f)}}{R_s + R_{in(f)}} = \dfrac{10 \times 10^{-3} \times 1302}{1000 + 1302} = 5.66 \ \text{mV}.$

Hence output voltage = $5.66 \times 60.15 = 340$ mV. (*Ans.*)

Note that the voltage gain of the amplifier before n.f.b. was applied was

$$A_v = A_i R_L / R_{in} = 164 \times 2200 / 6000 = 60.15$$

and this shows that voltage-shunt feedback reduces the current gain of an amplifier but does not affect its voltage gain.

Advantages of Negative Feedback

The application of any of the four types of negative feedback to an amplifier has the effect of reducing either the voltage gain or the current gain of that amplifier. This is obviously a disadvantage, but on the credit side, a number of desirable changes in the amplifier performance also take place. These are

(*a*) The gain stability is increased.
(*b*) Distortion and noise produced within the feedback loop *may* be reduced.
(*c*) The input and output impedances of the amplifier can be modified to almost any desired value.

Stability of Gain

For many applications it is important that an amplifier should have a more or less constant gain even though various parameters, such as power supply voltages and transistor parameters, may alter with time and/or with change in ambient temperature. The application of n.f.b. will considerably reduce the effect of such parameter variations on the overall gain of the amplifier.

If the loop gain, βA_v or βA_i, is much larger than unity the equations (3.2) and (3.15) reduce to

$$A_{v(f)} = A_{i(f)} = 1/\beta. \tag{3.21}$$

The gain of the amplifier is now merely a function of the components making up the feedback circuit, generally one or two resistors. Any changes in the parameters of the transistor, etc., will now have very little effect on the overall gain of the amplifier.

Example 3.6

An amplifier has a voltage gain before negative feedback is applied of 1200 ± 150. Negative feedback is to be applied to the amplifier to ensure that the gain will not vary by more than $\pm 1\%$. Calculate:

(*a*) the necessary feedback fraction β, and
(*b*) the voltage gain of the amplifier with the negative feedback applied.

Solution

$$1.01 A_{v(f)} = \frac{1350}{1+\beta 1350} \quad \text{and} \quad 0.99 A_{v(f)} = \frac{1050}{1+\beta 1050}.$$

Solving for β gives $\beta = 9.74 \times 10^{-3}$. (*Ans.*)

Then, $A_{v(f)max} = \dfrac{1350}{1+9.74 \times 10^{-3} \times 1350} = 95.4$ (*Ans.*)

$$A_{v(f)min} = \frac{1050}{1+9.74 \times 10^{-3} \times 1050} = 93.5. \quad (Ans.)$$

Amplitude—Frequency Distortion

Amplitude—frequency distortion of a complex signal is caused by the different frequency components of the signal being amplified by different amounts. The variation of gain with frequency of an amplifier was considered in Chapter 2 where it was shown to be caused by capacitive effects. The gain at low frequencies of an audio amplifier can be written as

$$A_{v(lf)} = \frac{A_{v(mf)}}{1-j\omega_1/\omega} \tag{2.18}$$

and the gain at high frequencies as

$$A_{v(hf)} = \frac{A_{v(mf)}}{1+j\omega/\omega_1}. \tag{2.24}$$

The expression for the overall low-frequency gain with feedback is now

$$A_{v(f)(lf)} = \frac{\dfrac{A_{v(mf)}}{1 - j\omega_1/\omega}}{1 + \dfrac{\beta A_{v(mf)}}{1 - j\omega_1/\omega}} = \frac{A_{v(mf)}}{1 + \beta A_{v(mf)} - j\omega_1/\omega}$$

$$= \frac{\dfrac{A_{v(mf)}}{1 + \beta A_{v(mf)}}}{1 - \dfrac{j\omega_1/\omega}{1 + \beta A_{v(mf)}}} = \frac{A_{v(f)(mf)}}{1 - \dfrac{j\omega_1/\omega}{1 + \beta A_{v(mf)}}}$$

The lower 3 dB frequency ω_1^1 now occurs when

$$1 = \frac{\omega_1}{\omega_1^1(1 + \beta A_{v(mf)})}$$

$$\omega_1^1 = \frac{\omega_1}{1 + \beta A_{v(mf)}}. \tag{3.22}$$

This is a lower frequency than that obtained without feedback.

Similarly, at high frequencies the overall gain with feedback is

$$A_{v(f)(hf)} = \frac{\dfrac{A_{v(mf)}}{1 + j\omega/\omega_2}}{1 + \dfrac{\beta A_{v(mf)}}{1 + j\omega/\omega_2}} = \frac{A_{v(f)(mf)}}{1 + \dfrac{j\omega/\omega_2}{1 + \beta A_{v(mf)}}}$$

and the upper 3 dB frequency ω_2^1 is

$$\omega_2^1 = \omega_2(1 + \beta A_{v(mf)}). \tag{3.23}$$

This simple analysis *indicates* that the bandwidth of an amplifier is increased in the same ratio as the gain is decreased by the application of n.f.b. However the results would only be true if the amplifier contained just the one time constant at low, or at high frequencies.

Figure 3.13 shows the gain–frequency characteristic of a typical amplifier before and after the application of n.f.b. Clearly, the feedback has made the gain of the amplifier much flatter over most of the frequency band.

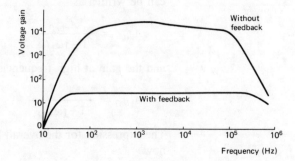

Fig. 3.13 Showing the effect of negative feedback on the gain-frequency characteristic of an amplifier.

Non-linearity Distortion

The mutual characteristics of all transistors exhibit some non-linearity and as a result the output and input signal waveforms are not identical. *Non-linearity distortion* is said to have occurred. The distortion exists because the output waveform contains components at frequencies that were not present in the input signal. If these new, unwanted frequencies are harmonically related to the input signal frequencies *harmonic distortion* is said to be present. If the new frequencies are equal to the sums and differences of frequencies combined in the input signal or harmonics of them, *intermodulation distortion* has taken place. Very often both harmonic and intermodulation distortion occur at the same time.

Suppose, for simplicity, that the input signal is sinusoidal and the distortion produced is due to a second-harmonic component only, say D.

The output of the amplifier is then

$$V_{out} = A_v V_x + D = A_v(V_{in} - \beta V_{out}) + D$$

$$V_{out}(1 + \beta A_v) = A_v V_{in} + D$$

$$V_{out} = \frac{A_v V_{in}}{1 + \beta A_v} + \frac{D}{1 + \beta A_v}. \tag{3.24}$$

According to equation (3.24) distortion is reduced by the same factor as the gain and so the percentage distortion is unchanged. Distortion is a function of the voltage level and is mainly produced in the final large-signal stage of an amplifier. For the same output level before and after the application of n.f.b., the input signal voltage with feedback must be increased $(1 + \beta A_v)$ times. Now the percentage distortion is reduced by the factor $(1 + \beta A_v)$. The extra signal must be provided by an earlier small-signal stage but this will introduce little, if any, distortion.

Equation (3.24) also assumes that the gain A_v of the amplifier is constant for all input signal levels; this of course is never true and, if it were, the gain characteristic would be linear and there would be no distortion in the first place. However, in practice, non-linearity distortion is nearly reduced by equation (3.24) provided it is not so severe that the gain before feedback is not approximately equal to the nominal value A_v at all levels.

Stability of Negative-feedback Amplifiers

When overall negative feedback is applied to a multi-stage amplifier, the fed-back voltage must be arranged to be in *antiphase* with the input signal voltage. At both low and high frequencies the phase shift through the amplifier will be altered because of the inevitable capacitive effects. If, at some particular frequency, the loop phase shift $\angle \beta A_v$ (or $\angle \beta A_i$) is 360° the feedback will become *positive*. There is then a possibility that the amplifier will oscillate unless the loop gain βA_v at this frequency is less than unity.

The problem is illustrated by the following example.

Example 3.7

A three-stage amplifier has three equal non-interacting 0.1 s time constants at low frequencies. If the mid-frequency voltage gain is $100 \angle 180°$, determine the greatest value of feedback factor β that can be used if the circuit is not to oscillate. If a greater value of β is used, calculate the frequency of the resulting oscillations.

Solution

The 'gain' of each *RC* circuit in the amplifier can be written in the form

$$A_{v(lf)} = \frac{A_{v(mf)}}{1-j/\omega\tau} = \frac{A_{v(mf)}}{1-j/0.1\omega}.$$

For three such identical stages,

$$\frac{A_{v(lf)}}{A_{v(mf)}} = \frac{1}{(1-j/0.1\omega)^3} = \left[\frac{1}{\sqrt{[1+(1/0.1\omega)^2]}}\right]^3 3\angle\tan^{-1}1/0.1\omega.$$

The amplifier introduces a 180° phase shift and for a loop phase shift of 360° the *RC* circuits must give a total phase shift of 180°.

Hence, the phase shift per circuit = 180°/3 = 60°, and

$$\tan^{-1}1/0.1\omega = 60°$$

$$\omega = 10/\tan 60° = 10/1.732 = 5.786 \text{ rad/s}.$$

Therefore $f = 5.786/2\pi = 0.92$ Hz. (*Ans.*)

At this frequency the magnitude of the gain, before feedback, of the amplifier is

$$\frac{100}{\{\sqrt{[1+(1/0.1\omega)^2]}\}^3} = \frac{100}{\{\sqrt{[1+(1.732)^2]}\}^3}$$

$$= 100/(\sqrt{4})^3 = 100/8.$$

The amplifier will oscillate if $|\beta A_v| = 1$ or if $\beta = 8/100$. (*Ans.*)

There are two main methods available by which the stability of a n.f.b. amplifier can be predicted. These two methods are the *Nyquist diagram* and the *Bode plot*.

The Nyquist Diagram

The loop gain of a feedback amplifier is a complex quantity, in that it has both amplitude and phase. The loop gain can be plotted on a graph using polar coordinates to produce the Nyquist diagram of the amplifier.

Nyquist's criterion states that: if the locus of βA_v does *not* enclose

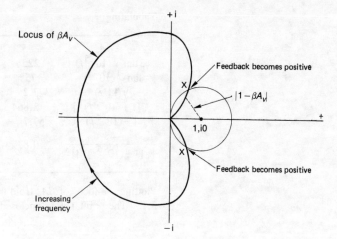

Fig. 3.14 Nyquist diagram of a feedback amplifier.

the point 1,j0 the circuit will be stable but, if the point 1,j0 is enclos-ed, the circuit will be unstable.

A circle of unity radius and centred upon the point 1,j0 can be drawn on the same axes, and then whenever the locus of βA_v is *inside* this circle the feedback is *positive*; conversely, whenever the locus of βA_v is *outside* the circle the feedback is *negative*. Consider the typical Nyquist diagram shown in Fig. 3.14.

It can be seen that the point 1,j0 is not enclosed by the locus of βA_v and this means that the amplifier is stable, i.e. it will not oscillate. The locus does enter the unity circle at the points XX and so the feedback becomes positive at both low and high frequencies. When the locus of βA_v is inside the circle, it is clear that the line $|1-\beta A_v|$ must have a length less than 1 and so

$$|A_{v(f)}| = \frac{A_v}{|1-\beta A_v|}$$

must be greater than A_v and the feedback is positive.

Example 3.8

An amplifier has a voltage gain before feedback given by

$$A_v = \frac{-1200}{(1+jf\times10^{-5})^3}$$

and a feedback factor of $0.04 \angle 0°$. Draw the Nyquist diagram for the amplifier and use it to determine whether or not the amplifier is stable.

Solution

$$\beta A_v = \frac{-1200\times0.04}{(1+jf\times10^{-5})^3} = \frac{48\angle180°}{\{\sqrt{[1+(10^{-5}f)^2]}\}^3\,3\tan^{-1}(10^{-5}f)}$$

Tabulating:

$\angle \beta A_v$	$-45°$	$0°$	$45°$	$90°$	$135°$	$180°$
$3\tan^{-1}(10^{-5}f)$	$225°$	$180°$	$135°$	$90°$	$45°$	$0°$
$\tan^{-1}(10^{-5}f)$	$75°$	$60°$	$45°$	$30°$	$15°$	$0°$
f(kHz)	373.2	173.2	100	57.7	26.8	0
$\sqrt{[1+(10^{-5}f)^2]}$	3.864	2.0	1.414	1.155	1.035	1
$\{\sqrt{[1+(10^{-5}f)^2]}\}^3$	57.7	8.0	2.828	1.539	1.11	1
$\dfrac{48}{\{\sqrt{[1+(10^{-5}f)^2]}\}^3}$	0.83	6	16.97	31.19	43.26	48

Plotting the locus of βA_v from these figures gives the Nyquist plot of Fig. 3.15. The point 1,j0 *is* enclosed and so the amplifier is unstable.

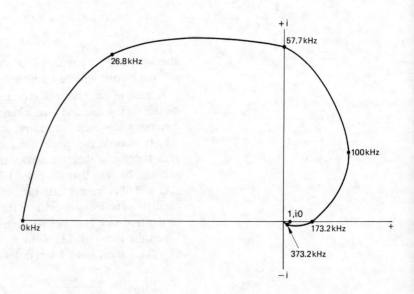

Fig. 3.15

Gain and Phase Margin

If an amplifier is shown by its Nyquist diagram to be stable it is of considerable importance to be able to determine how near the circuit is to instability. There are two ways in which the relative stability of an amplifier can be specified, namely the *gain margin* and the *phase margin*.

The *gain margin* of an amplifier is the amount by which the gain must be increased in order to produce instability. Generally, the gain margin is expressed in decibels. Thus, referring to Fig. 3.16(a), the βA locus crosses the real axis at a frequency f_1 at the point 0.5,j0. The increase in the magnitude of the loop gain $|\beta A_v|$ needed to produce instability, i.e. for the locus to reach the point 1,j0, is 1/0.5 or 2. Thus, the gain margin is $20\log_{10}2$ or 6 dB.

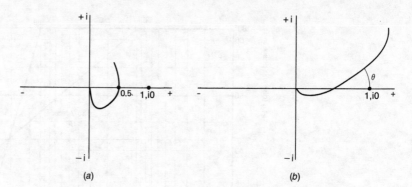

Fig. 3.16 (*a*) Gain margin, (*b*) phase margin.

(*a*)   (*b*)

The *phase margin* is $\angle \beta A_v$ at the frequency at which $|\beta A_v|$ is unity. It is the amount by which the phase of the loop gain must be altered to make the amplifier unstable. The phase margin is shown as θ in Fig. 3.16(*b*).

It is generally reckoned that for good stability an amplifier should have a gain margin of at least 14 dB (a factor of 5) and a phase margin of at least 45°. If the gain margin is smaller than this, undesirable peaks will appear in the gain–frequency characteristic of the amplifier.

The gain margin specification is not appropriate for an amplifier in which the angle of the loop phase $\angle \beta A_v$ never reaches 180°; for such an amplifier the phase margin should be quoted.

Bode Plots

The Bode plot of an amplifier consists of two graphs in which the amplitude *in dB* and the phase of the gain are plotted separately to a base of frequency *plotted to a logarithmic scale* (see pages 59 and 60).

Consider, as an example, the amplifier which was the subject of Example 3.8, i.e.

$$A_v = \frac{-1200}{(1+jf10^{-5})^3} \quad \text{and} \quad \beta = 0.04 \angle 0°.$$

The amplifier will oscillate if $\beta A_v = 1 \angle 180°$

$$|A_v| \, \text{dB} = 20\log_{10} 1200 - 60\log_{10} \sqrt{[1+f^2 \times 10^{-10}]}$$

$$\simeq 61.6 \, \text{dB} - 60\log_{10} 10^{-5}f.$$

The 3 dB frequency is 100 kHz and so

$$f_2/10 = 10 \, \text{kHz} \quad \text{and} \quad 10f = 1 \, \text{MHz}.$$

The Bode diagram for this amplifier is shown in Fig. 3.17. From the figure, $\theta = 360°$ when the frequency is approximately 205 kHz and $A_v = 42$ dB. For oscillations to occur $|\beta A_{v(hf)}| = 1$ or

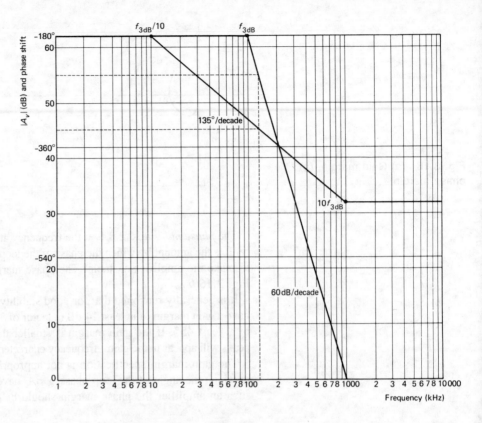

Fig. 3.17

$$20\log_{10}|A_{v(hf)}| + 20\log_{10}|\beta| = 0.$$

Hence $\quad 20\log_{10}|A_{v(hf)}| = 20_{10}\log|1/\beta|.$

So for oscillations $|1/\beta| = |A_{v(hf)}| = 42$ dB $= 126$. This is greater than the actual value of β, i.e. $1/\beta = 1/0.04 = 25$ and so the amplifier is unstable.

For stability, the magnitude of the gain must be made to fall at a faster rate to ensure that the loop gain is less than unity at 205 kHz.

If a particular phase margin is required, say 30°, then $\angle\beta A = 360° - 30° = 330°$ which occurs at a frequency of 130 kHz when $|A_v| = 55$ dB. The phase margin is defined for $|\beta A_v| = 1$ or 0 dB and hence $|1/\beta| = |A_v|$ to achieve the required margin. In this case the required β value is 1.78×10^{-3}. A similar analysis can be applied to the low-frequency end of the gain characteristic.

An amplifier cannot become unstable with only two time constants and this means that low-frequency stability can probably be assured by the use of direct coupling between stages.

High-frequency stability can be achieved by the use of *phase-lead compensation*. This consists of the connection of a capacitor C_f in parallel with the feedback resistor R_f as shown in Fig. 3.18.

From equation (3.19), $\beta = R_L/(R_f+R) \simeq R_L/R_f$. At high frequencies where the reactance of C_f is low enough to affect the response of the amplifier $\beta_{(hf)} = R_L/Z_f$ where

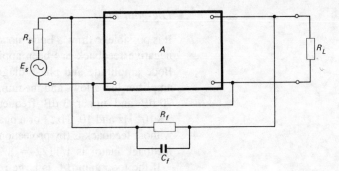

Fig. 3.18 Phase-lead compensation.

$$Z_f = \frac{R_f}{1 + j\omega C_f R_f} = \frac{R_f}{1 + j\omega/\omega_2}.$$

Therefore,

$$\beta_{(hf)} = \frac{R_L}{R_f}(1 + j\omega/\omega_2) = \beta(1 + j\omega/\omega_2) = \beta(1 + j\omega\tau).$$

The high-frequency gain of an amplifier with three h.f. time constants is of the form

$$A_{v(hf)} = \frac{A_{v(mf)}}{(1 + j\omega\tau_a)(1 + j\omega\tau_b)(1 + j\omega\tau_c)}$$

where τ_a, τ_b and τ_c are the three h.f. time constants. The gain with feedback is

$$A_{v(hf)} = \frac{A_{v(mf)}/(1 + j\omega\tau_a)(1 + j\omega\tau_b)(1 + j\omega\tau_c)}{1 + \dfrac{\beta(1 + j\omega\tau)A_{v(mf)}}{(1 + j\omega\tau_a)(1 + j\omega\tau_b)(1 + j\omega\tau_c)}}.$$

If τ is chosen to be equal to one of the three amplifier time constants, one term in the denominator of $A_{v(hf)}$ will be cancelled out. The loop gain then has the high-frequency phase characteristic of a two-stage amplifier and will be stable.

Example 3.9

An amplifier has three high-frequency time constants having 3 dB frequencies of, respectively, 10^5 Hz, 2×10^5 Hz and 10^6 Hz. Phase-lead compensation is to be used to cancel the 3 dB frequency at 2×10^5 Hz. Calculate the value of capacitance needed. The feedback resistor is 200 kΩ.

Solution

$$C_f = \frac{1}{2\pi f R_f} = \frac{1}{2\pi \times 2 \times 10^5 \times 2 \times 10^5} = 3.98 \text{ pF.} \quad (Ans.)$$

Log-gain Plots

It is possible to draw a Bode plot of an amplifier's characteristics *after* negative feedback has been applied. Consider the amplifier whose Bode amplitude and phase diagrams are shown in Fig. 3.19. The amplifier has a low- and medium-frequency current gain of 3162 or 70 dB and upper 3 dB frequencies or *break points* at 10^4 Hz, 3×10^4 Hz and 10^5 Hz. For a phase margin of 45° the amplifier gain without feedback is (by projection) 61 dB or 1122. Thus, the required feedback factor is $1/1122 = 8.9 \times 10^{-4}$.

If the loop gain βA_v is large relative to unity, the gain with feedback is approximately equal to $1/\beta$ or 1122. This is the same value

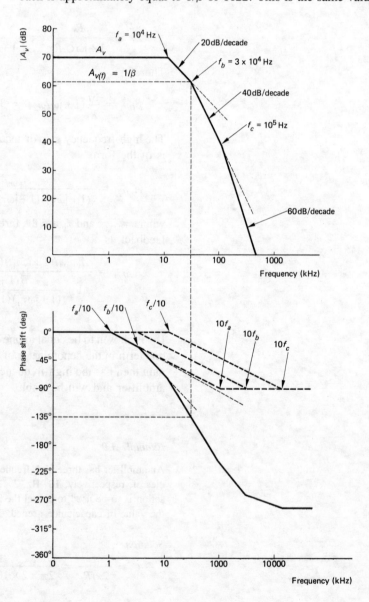

Fig. 3.19 Log-gain plot of an amplifier.

as the gain *without feedback* at the frequency at which the phase margin is 45° (or whatever phase margin is wanted). This means that the *closed loop* and the *open loop* gain lines intersect at the frequency at which the loop gain is unity. This is shown by the line labelled $A_{v(f)} = 1/\beta$. For stability with 45° phase margin, this intersection must take place at a frequency at or below the second break frequency. At higher frequencies the loop gain βA_v is less than unity and the open- and closed-loop gains are approximately equal — absolutely equal on the straight line approximation of the Bode diagram.

The upper 3 dB frequency of the amplifier with feedback is very nearly the first break frequency of the closed-loop curve. This, of course, occurs at the frequency at which the closed-loop and the open-loop curves intersect.

The effect of phase-lead compensation is shown in Fig. 3.20. The gain with feedback is approximately

$$1/\beta_{(hf)} = 1/\beta(1+j\omega\tau)$$

and it therefore has a break frequency at $f = 1/2\pi\tau$. The figure has been drawn assuming that $f = 3 \times 10^4$ Hz so that the middle break point of the open loop gain is cancelled. Now the A_v and $A_{v(f)}$ lines intersect at the highest break frequency of 10^5 Hz, and the upper 3 dB frequency of the compensated amplifier is 3×10^4 Hz. The difference in the slopes of A_v and $A_{v(f)}$ at the point of intersection must not be greater than 20 dB/decade and this means that $A_{v(f)}$ must be lower when phase-lead compensation is used than if the amplifier is uncompensated.

An alternative method of improving the stability of a feedback amplifier is to apply extra shunt capacitance to one of the stages and so reduce its upper 3 dB frequency. The magnitude of the loop gain becomes less than unity at a frequency where the amplifier introduces negligible phase shift. For stabililty, the line representing $A_{v(f)} = 1/\beta$

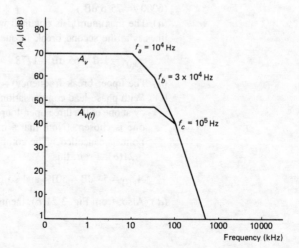

Fig. 3.20 Showing the effect of phase-lead compensation.

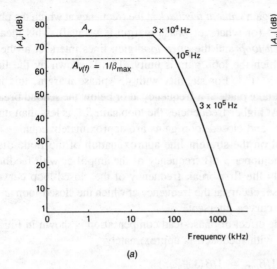

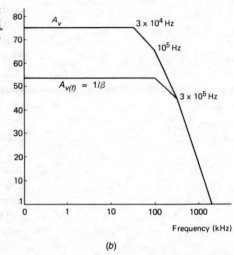

Fig. 3.21

(a)

(b)

should intersect the A_v line at or below the second break frequency, and clearly this is made easier if the lowest break frequency f_a is shifted to the left (Fig. 3.19).

Example 3.10

An amplifier has a mid-frequency gain before feedback of 6000 and upper break frequencies of 3×10^4 Hz, 10^5 Hz and $3 \cdot 10^5$ Hz respectively. Calculate (a) the maximum β possible for 45° phase margin if compensation is not applied, (b) the upper break frequency of the feedback amplifier, (c) the maximum β possible if phase-lead compensation is applied with a break frequency at 10^5 Hz, (d) the voltage gain and the upper break frequency with phase-lead compensation.

Solution

The Bode diagram for the uncompensated amplifier is shown by Fig. 3.21(a). (6000 = 75.6 dB.)

(a) The maximum β is obtained when the intersection of the A_v and $A_{v(f)}$ lines is at the second break frequency. Hence, from the figure

$$A_{v(f)} = 1/\beta = 65 \text{ dB} = 1778 \quad \text{and so } \beta_{max} = 5.62 \times 10^{-4}. \quad (Ans.)$$

(b) The upper break frequency = 10^5 Hz. (Ans.)

(c) With phase-lead compensation, the $A_v = 1/\beta$ line has a break frequency at one of the three open-loop break frequencies but usually the middle one is chosen. (Note that a higher β is possible if the highest break point is cancelled.) The compensated Bode diagram is shown in Fig. 3.21(b). From this

$$A_{v(f)} = 54 \text{ dB} = 501 \quad \text{and so } \beta = 1/501 = 2 \times 10^{-3}. \quad (Ans.)$$

(d) Also, from Fig. 3.21(b), the upper break frequency is 10^5 Hz. (Ans.)

4 Operational Amplifiers

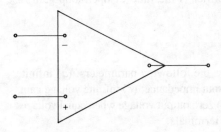

Fig. 4.1 Op-amp symbol.

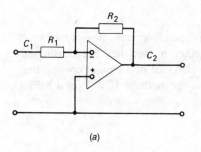

(a)

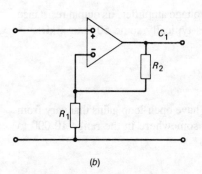

(b)

Fig. 4.2 Op-amp connected to provide (a) an inverting gain and (b) a non-inverting gain.

An operational amplifier, or *op-amp*, is a monolithic integrated circuit amplifier which has a very high voltage gain, a high input impedance, and (for most types) a low output impedance. These are, of course, the desirable parameters of a voltage amplifier and they make the op-amp suitable for use in a wide variety of applications. A few types are known as operational transconductance amplifiers (o.t.a.) and these have a high output impedance. A monolithic op-amp is essentially a d.c. differential amplifier of high gain, and the symbol for the device is shown in Fig. 4.1.

Two input terminals are provided. One of them, labelled −, is known as the inverting terminal since a voltage applied to this terminal appears at the output terminal with the opposite polarity, i.e. a sinusoidal input signal will experience a phase shift of 180°. The other input terminal is labelled + and this is the non-inverting terminal; a signal applied to it is amplified with zero phase shift. Only a very small difference in potential between the two inputs is needed to produce a large output voltage. When this difference is equal to zero, the output voltage should also be zero but, for reasons to be explained later, this is not always the case. The voltage gain is large and unpredictable and is *always* reduced by negative feedback, when the device is used as an amplifier. Two further terminals are generally provided for the connection of positive and negative power supply voltages. Two polarities are necessary so that the output voltage can vary either side of zero volts, although some types of op-amp are able to operate from a single positive supply voltage. Most types of op-amp will operate satisfactorily from a wide range of supply voltages.

Most op-amps employ bipolar transistors throughout their internal circuitry. An increased input resistance and reduced input bias current are possible if the input stage uses field effect transistors. Op-amps with a junction FET input stage are known as BiFET devices while another technology that uses MOSFETs in the input stage is known as BiMOS. Also available are op-amps which employ CMOS circuitry throughout.

There are two ways in which an operational amplifier can be connected to act as an amplifier; one method gives an inverting gain and the other gives a non-inverting gain. The connections are shown in Figs. 4.2(a) and (b)

99

The voltage gains of the two circuits are given by equations (4.1) and (4.2), respectively.

$$A_{v(f)} = \frac{V_{out}}{V_{in}} = \frac{-R_2}{R_1} \tag{4.1}$$

and

$$A_{v(f)} = \frac{R_1 + R_2}{R_1}. \tag{4.2}$$

The loop gain of either amplifier is the ratio of open-loop gain A_v to closed-loop gain $A_{v(f)}$.

Parameters of Operational Amplifiers The ideal op-amp would have the following parameters: (*a*) infinite input impedance; (*b*) zero output impedance; (*c*) infinite voltage gain; (*d*) infinite bandwidth; and (*e*) zero output voltage when equal voltages are applied to its two input terminals.

In practice, of course, the ideal op-amp does not exist and practical devices have various limitations.

Input Resistance

The input resistance of an op-amp depends upon the type of transistors used in its input stage and whether or not they are Darlington-connected. Typically, the input resistance of a bipolar op-amp may be in the range 100 kΩ to 10 MΩ and perhaps 10^{12} Ω for a BiFET or BiMOS type.

Output Resistance

Since the op-amp is essentially a voltage amplifier, its output resistance should be low as possible and is typically some 50 Ω to 4 kΩ.

Voltage Gain

Commercially available op-amps have open-loop gains that vary from one type to another but may be somewhere in the range 10 000 to 200 000.

Input Bias and Offset Currents

The input stage of an op-amp is always a differential amplifier and both of its transistors must be correctly biased. Hence a small base bias current is taken by each of the input transistors. The *bias current* I_{B-} taken by the inverting input will usually *not* be of exactly the same value as the bias current I_{B+} that flows into the non-inverting terminal. The data sheet of an op-amp generally specifies the average input bias current I_B, where

$$I_B = \frac{I_{B-} + I_{B+}}{2}. \tag{4.3}$$

The value of the input bias current I_B may range from a few picoamps for a BiFET, BiMOS or CMOS device to a few microamps for a bipolar device.

The *input offset current I_{OS}* is equal to the difference between the input bias currents I_{B-} and I_{B+}. Thus

$$I_{OS} = I_{B-} - I_{B+}. \tag{4.4}$$

Usually, I_{OS} will be somewhere between 10% and 25% of the input bias current.

Effect of Input Bias Current on Output Voltage

The input bias currents I_{B-} and I_{B+} may flow through external resistances connected to the op-amp and develop a d.c. voltage that may produce an unwanted *offset voltage* at the output. Figure 4.3 shows an op-amp connected as an inverting amplifier with an input voltage V_{IN} and an output voltage V_{OUT}. V_- and V_+ are, respectively, the voltages between earth and the inverting, and the non-inverting, terminals; in this case $V_+ = 0$.

Summing the currents at the inverting terminal,

$$\frac{V_{IN} - V_-}{R_1} + \frac{V_{OUT} - V_-}{R_2} + I_{B-} = 0.$$

Since the open-loop gain of the op-amp is high V_- is approximately equal to V_+ or zero, and hence

$$\frac{V_{IN}}{R_1} + \frac{V_{OUT}}{R_2} + I_{B-} = 0 \tag{4.5}$$

and

$$V_{OUT} = \frac{-V_{IN} R_2}{R_1} - I_{B-} R_2. \tag{4.6}$$

The term $I_{B-} R_2$ represents an unwanted component of the output voltage which could be reduced by using a small resistance value for the feedback resistor R_2. This step would however reduce the voltage gain obtainable for a given input resistance (set by R_1). A better alternative is to connect a third resistor R_3 into the circuit between the non-inverting terminal and earth, see Fig. 4.4. Now the voltage between the positive terminal and earth is no longer zero but is equal to $I_{B+} R_3$ so that $V_- \simeq V_+ \simeq I_{B+} R_3$ and equation (4.5) becomes

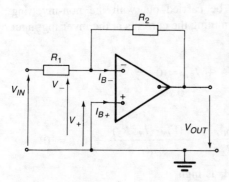

Fig. 4.3 Effect of bias currents in inverting amplifier.

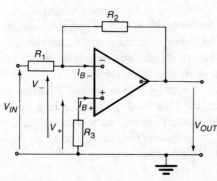

Fig. 4.4 Reducing the effect of bias currents.

$$\frac{V_{IN}-I_{B+}R_3}{R_1} + \frac{V_{OUT}-I_{B+}R_3}{R_2} + I_{B-} = 0$$

and

$$V_{OUT} = \frac{-V_{IN}R_2}{R_1} + I_{B+}R_3\left(1+\frac{R_2}{R_1}\right) - I_{B-}R_2. \qquad (4.7)$$

The second and third terms in equation (4.7) are both unwanted components of the output voltage but their effect can be minimized by arranging that $R_2 = R_3(1+R_2/R_1)$. Then, $R_3 = R_1R_2/(R_1+R_2)$ and the output offset voltage is equal to $R_2(I_{B-}-I_{B+})$ or $I_{OS}R_2$. Thus, the output offset voltage produced by the input bias currents can be given a minimum value that is equal to the product of the feedback resistance and the input offset current.

The compensating resistor R_3 should have a resistance value equal to the resistance of R_1 and R_2 in parallel. Any source resistance should be included in this calculation by adding its value to the resistance of R_3.

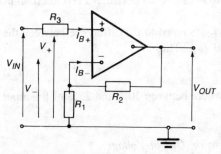

Fig. 4.5 Effect of bias currents in non-inverting amplifier.

A similar analysis can be carried out with the non-inverting amplifier, see Fig. 4.5. Summing the currents at the inverting input terminal,

$$\frac{-V_-}{R_1} + \frac{V_{OUT}-V_-}{R_2} + I_{B-} = 0$$

Therefore

$$\frac{-V_{IN}+I_{B+}R_3}{R_1} + \frac{V_{OUT}}{R_2} - \frac{(V_{IN}-I_{B+}R_3)}{R_2} + I_{B-} = 0$$

Since the open-loop gain A_v is high

$$V_+ = V_{IN}-I_{B+}R_3 \simeq V_-$$

and so,

$$V_{OUT} = V_{IN}(1+R_2/R_1) + I_{B+}R_3(1+R_2/R_1) - I_{B-}R_3. \qquad (4.8)$$

The output voltage consists of the sum of the wanted output of the non-inverting amplifier plus the *same* offset terms as were obtained for the inverting amplifier. As before, the minimum offset voltage will be obtained if the resistance of R_3 plus any source resistance is made equal to the resistance of R_1 and R_2 in parallel.

The inclusion of R_3 in Figs. 4.4 and 4.5 to overcome the effects of the input bias current suffers from the disadvantage that the source resistance must be of fixed and constant value. If the source resistance is a variable, a more complex arrangement is necessary and Fig. 4.6 shows one of the simplest of the possible circuits.

Input Offset Voltage

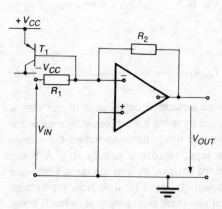

Fig. 4.6 Bias current compensation.

The output voltage of an op-amp may not be at zero volts even when there is no input voltage because the input terminals have been shorted

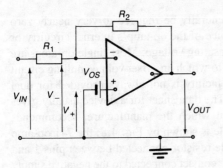

Fig. 4.7 Input offset voltage.

together, *and* there are zero input bias current effects. The unwanted output offset voltage is the result of various unbalances that exist within the internal circuitry of the op-amp. It is convenient to suppose that the unwanted offset voltage is the result of an *input offset voltage V_{OS}* that is applied in series with either one of the input terminals. Consider Fig. 4.7 in which the input offset voltage is represented by a battery connected in series with the non-inverting terminal. Summing the currents at the inverting input,

$$\frac{V_{IN}-V_-}{R_1} + \frac{V_{OUT}-V_+}{R_2} = 0.$$

Since the open-loop voltage gain of the op-amp is very high $V_- \simeq V_+ = V_{OS}$. Therefore,

$$\frac{V_{IN}-V_{OS}}{R_1} + \frac{V_{OUT}-V_{OS}}{R_2} = 0$$

or

$$V_{OUT} = \frac{-V_{IN}R_2}{R_1} + V_{OS}(1+R_2/R_1). \tag{4.9}$$

Typical and maximum values for V_{OS} are usually quoted in the data sheet of an op-amp.

Example 4.1

An op-amp has input bias currents of $I_{B+} = 0.5\ \mu A$, $I_{B-} = 0.4\ \mu A$ and an input offset voltage of 1 mV. It is used in the circuit shown in Fig. 4.4. If the gain of the circuit is 20 and its input resistance is 1000 Ω, calculate (*a*) the optimum value for R_3, and (*b*) the total offset output voltage when (*a*) is employed.

Solution
Since the input resistance of 1000 Ω is set by R_1, R_2 must be equal to 20 kΩ.

(*a*) $R_3 = \left(\dfrac{20 \times 1}{20+1}\right)$ kΩ = 952 Ω. (*Ans.*)

(*b*) Offset output voltage = $0.1 \times 10^{-6} \times 20 \times 10^3 + 1 \times 10^{-3}(1+20)$

$$= 2\ \text{mV} + 21\ \text{mV} = 23\ \text{mV}. \quad (Ans.)$$

If the non-inverting amplifier, Fig. 4.5, is considered assume V_{OS} to be in series with the positive terminal. Then, since the gain is very high, $V_- = V_+ = V_{IN}+V_{OS}$. Equating the currents in R_1 and R_2 gives,

$$\frac{V_{IN}+V_{OS}}{R_1} = \frac{V_{OUT}-V_{IN}-V_{OS}}{R_2}$$

$$V_{OUT} = V_{IN}R_2\left(\frac{1}{R_1}+\frac{1}{R_2}\right) + V_{OS}R_2\left(\frac{1}{R_1}+\frac{1}{R_2}\right)$$

$$= (1+R_2/R_1)V_{IN} + (1+R_2/R_1)V_{OS}. \tag{4.10}$$

The offset voltage can generally be reduced to very nearly zero by either adjusting the balance of the op-amp's internal circuitry or by the insertion of a compensating voltage. Most single op-amps are provided with package pins to which an offset voltage nulling circuit can be connected, but this facility is not always provided for dual or quad op-amp packages. The data sheet for a device usually gives details of the nulling circuit which the manufacturer recommends should be used. One example is shown by Fig. 4.8; the 741 op-amp should have a 10 kΩ variable resistor connected between pins 1 and 5, with the resistor's variable contact connected to the negative supply voltage. Other op-amps employ similar circuits but generally with different resistance values.

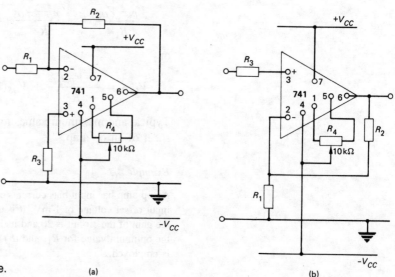

Fig. 4.8 Minimizing offset voltage. (a) (b)

Drift

Both the input bias currents and the input offset voltage are temperature dependent and any changes that occur for this reason are known as *drift*. Typical figures are given in data sheets, often in graphical form, as a temperature coefficient quoted in either nA/°C or in μV/°C.

Example 4.2

The 301 op-amp has a maximum I_{OS} drift of 0.3 nA/°C and a maximum V_{OS} drift of 30 μV/°C over the temperature range of 25 °C to 75 °C. It is used in the circuit shown in Fig. 4.4 with $R_1 = 10$ kΩ and $R_2 = 100$ kΩ. If the circuit has been set for zero offset voltage at 25 °C by a nulling circuit calculate the total output offset voltage at 50 °C.

Solution

Change in $I_{OS} = 0.3 \times 10^{-9} \times 25 = 7.5$ nA.

Therefore the change in the output offset voltage due to this cause is $7.5 \times 19^{-9} \times 100 \times 10^3 = 0.75$ mV.

Change in $V_{OS} = 30 \times 10^{-6} \times 25 = 0.75$ mV.

Therefore the total change in the output offset voltage, which is now the actual offset voltage, is $0.75 + 0.75 \times 11 = 9$ mV.   (*Ans.*)

The effects of the input bias currents and of the input offset voltage have been treated separately to simplify matters but, in practice, their effects are always simultaneously present.

If operation at, or near, zero frequency is not required, the effects of any output offset voltage can be eliminated by the use of an output coupling capacitor and the other precautions may not be necessary. Care must still be taken to ensure that the offset is not large enough to cause the op-amp to saturate on the peaks of the output signal voltage. Also, of course, the capacitor will cause the gain of the circuit to fall at lower frequencies, as discussed in Chapter 2.

Bandwidth

When an op-amp is used to amplify small-signal a.c. voltages that produce an output voltage of no more than about 1 V its *gain-frequency characteristic* will be its most important parameter. For larger signals another parameter, known as the *slew rate*, may be of greater significance.

Since all op-amps are operated within a feedback loop there always exists a potential risk of instability and oscillation at higher frequencies. Many op-amps are *internally compensated* to avoid instability; this means that an internally connected capacitor is used to make the voltage gain fall off with increase in frequency.

The *small-signal response* curve of an op-amp illustrates how its open-loop gain varies with frequency, and Fig. 4.9 shows the curve

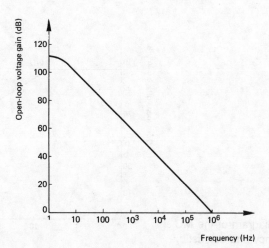

Fig. 4.9 Open-loop response of a 741 op-amp.

for the 741 op-amp. At low frequencies (up to about 7 Hz) the large-signal gain is, typically, equal to 200 000. The first *break point* or *break frequency* occurs when the gain has fallen by 3 dB from its low-frequency value. At frequencies higher than the break frequency the gain falls at a constant rate of 6 dB/octave (or 20 dB/decade) and eventually reaches *unity* at about 1 MHz.

The frequency at which the op-amp has an open-loop voltage gain of unity is known as the *small-signal unity-gain bandwidth*. It is also sometimes called the *gain-bandwidth product*. The unity-gain bandwidth may be specified in the data sheet, either by quoting a figure or graphically, or it may be calculated from the *transient-response risetime* which may alternatively be specified. For the 741 the transient-response risetime is typically equal to 0.3 μs. The unity-gain bandwidth can be calculated using equation (4.11), i.e.

$$\text{unity-gain bandwidth} = \frac{0.35}{\text{transient-response risetime}}. \quad (4.11)$$

Thus, for the 741 the unity-gain bandwidth is $0.35 \times 10^6/0.3$ or 1.167 MHz $\simeq$ 1 MHz. Since the open-loop gain-frequency characteristic of a single-pole op-amp falls at 6 dB/octave above the break frequency the unity-gain bandwidth can be used to find the gain provided at any other frequency. Again, for the 741: the open-loop gain at 100 kHz is 10, at 10 kHz is 100, and so on.

Not all op-amps exhibit a single break frequency in their open-loop frequency response: some have two, or more, breakpoints. One example, the gain-frequency characteristic of the MC 1539 is shown in Fig. 4.10 in Bode plot form; the response includes four break points at approximately 5 kHz, 300 kHz, 1 MHz, and 2 MHz. Op-amps that are not internally compensated need to be given external compensation before they can be used, and then their gain-frequency response will depend upon the values of the compensating components chosen. The data sheet for the 101/201/301 op-amp (p. 112) shows this clearly.

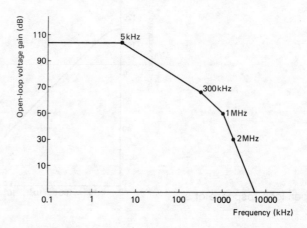

Fig. 4.10 Bode diagram of the MC 1539 gain–frequency characteristic.

Slew Rate

The slew rate of an op-amp is the maximum rate, in $V/\mu s$, at which its output voltage is capable of changing when the maximum output voltage is being supplied, i.e.

$$\text{Slew rate} = \frac{\mathrm{d}V_{out(max)}}{\mathrm{d}t} \; V/\mu s. \qquad (4.12)$$

The op-amp's internal circuitry includes a capacitor across which the voltage must change before the output voltage will be able so to do. Since $i = C\,\mathrm{d}v/\mathrm{d}t$, a high rate of change of voltage requires a high current/capacitance ratio, and the slew rate is equal to the ratio (maximum current I the capacitor can pass)/(capacitance C). For the 741, $I = 15\ \mu A$ and $C = 30\ pF$ to give a slew rate of $(15 \times 10^{-6})/(30 \times 10^{-9}) = 0.5\ V/\mu s$. This is a much lower figure than many other op-amps are able to offer and slew rates of 20 $V/\mu s$ or more are available.

When a signal at a given frequency is applied to an op-amp, the maximum permissible output voltage is determined by the slew rate; should a greater output voltage be developed, the signal waveform will be distorted. This is shown by Fig. 4.11.

The *full-power bandwidth* of an op-amp is the highest-frequency full-voltage sinusoidal signal that can appear at the output without slew-rate limitations causing waveform distortion.

The slew rate and the full power bandwidth are related since

$$v_{out(max)} = V_{out(max)} \sin 2\pi f_m t$$

where f_m is the full-power bandwidth upper frequency limit in Hz. Therefore,

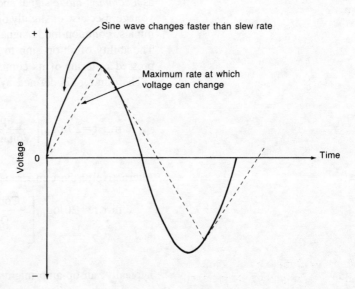

Fig. 4.11 Waveform distortion caused by inadequate slew rate.

$$\frac{\mathrm{d}v_{out(max)}}{\mathrm{d}t} = 2\pi f_m V_{out(max)} \cos 2\pi f_m t$$

and since the maximum rate of change occurs as the waveform crosses the zero axis then

$$\text{Slew rate} = 2\pi f_m V_{out(max)}. \tag{4.13}$$

Equation (4.13) illustrates that, if the maximum peak output voltage swing is reduced, the operating frequency can be increased, and vice versa.

Example 4.3

An op-amp has a maximum output voltage of 12 V and a slew rate of 5 V/μs. Calculate its full power bandwidth.

Solution
From equation (4.13)

$$f_m = \text{slew rate}/2\pi V_{out} = 5 \times 10^6/(2\pi \times 12) = 66.3 \text{ kHz}. \quad (Ans.)$$

The maximum output voltage $V_{out(max)}$ of an op-amp is always somewhat less than the supply voltages $\pm V_{CC}$ applied to the IC. For the 741, if $V_{CC} = \pm 15$ V then $V_{out(max)} = \pm 14$ V, but if $V_{CC} = \pm 22$ V then $V_{out(max)} = \pm 16$ V.

Common-mode Rejection Ratio

The output voltage of an op-amp is proportional to the *difference* between the voltages applied to its inverting and non-inverting terminals. When these two voltages are equal, the output should ideally be zero. A signal that is simultaneously applied to both input terminals is known as a *common-mode* signal and it is nearly always an unwanted noise voltage. Because of slightly different gains between the two input terminals, common-mode signals do *not* entirely cancel at the output. The ability of an op-amp to suppress common-mode signals is expressed in terms of its common-mode rejection ratio (c.m.r.r.).

The c.m.r.r. is defined by

$$\text{c.m.r.r} = 20\log_{10}\left[\frac{\begin{array}{c}\text{Voltage gain for signal}\\ \text{applied to} + \text{ or to} - \text{ terminal}\end{array}}{\text{Voltage gain for common-mode signal}}\right] \text{dB.} \tag{4.14}$$

Alternatively, the c.m.r.r. is often expressed thus:

$$\text{c.m.r.r.} = 20\log_{10}\left[\frac{\text{Common-mode input voltage}}{\begin{array}{c}\text{Differential input voltage}\\ \text{for the same output}\end{array}}\right] \text{dB.} \tag{4.15}$$

Typically, an op-amp might have a c.m.r.r. of 90 dB.

Example 4.4

An op-amp has an inverting gain of 150 000 and a non-inverting gain of 149 960. Calculate its c.m.r.r.

Solution

When a common-mode signal is applied

$$V_{out(cm)} = (-150\ 000 + 149\ 960) V_{in(cm)} = -40 V_{in(cm)}.$$

Therefore, the common-mode gain is 40.

The average gain is $(150\ 000 + 149\ 960)/2 = 149\ 980$.

Hence the c.m.r.r. $= 20 \log_{10}[149\ 980/40] = 71.5$ dB. (*Ans.*)

Data Sheets

Information about the package, the pin connections, the current and voltage ratings, and the parameters of an op-amp can all be obtained from the data sheet which is made available by the manufacturer. A data sheet is headed by the type number of the op-amp and a brief statement of its nature and main intended field of operation. Then follow schematic and connection diagrams, absolute maximum ratings and electrical characteristics. Some data sheets, although not all, then provide a series of graphs that give data on various performance characteristics.

As examples, the data sheets for the LM741 and the LM 101/201/301 op-amps are given in Figs. 4.12 and 4.13, respectively.

Semiconductor distributors' catalogues generally provide concise data in tabular form on a number of op-amps. Table 4.1 shows, as an example, the data provided by the RS Components' catalogue (Courtesy of RS Components Ltd). The table can be used to compare the various op-amps with regard to their desirable characteristics and hence to make a short list of suitable devices. Once this has been made, the data sheets of the possible op-amps should be obtained and scrutinized to enable the choice for the particular application in mind to be made.

Internal Circuitry

Both the 741 and the 101/201/301 op-amps, and indeed most other types as well, have the internal structure that is shown in block diagram form in Fig. 4.14. Only the first two blocks make a contribution to the overall open-loop gain of the op-amp. The first, input, stage is always a differential amplifier because this gives the op-amp two input terminals, makes decoupling capacitors unnecessary, and provides both low drift and high c.m.r.r. The majority of op-amps employ bipolar transistors in this stage but some employ either junction FETs or MOSFETs. Extra amplification is given by the gain stage. The level shifting stage is necessary to set the quiescent output voltage to 0 V and to ensure that the output voltage will be able to go either

National Semiconductor

Operational Amplifiers/Buffers

LM741/LM741A/LM741C/LM741E Operational Amplifier

General Description

The LM741 series are general purpose operational amplifiers which feature improved performance over industry standards like the LM709. They are direct, plug-in replacements for the 709C, LM201, MC1439 and 748 in most applications.

The amplifiers offer many features which make their application nearly foolproof: overload protection on the input and output, no latch-up when the common mode range is exceeded, as well as freedom from oscillations.

The LM741C/LM741E are identical to the LM741/LM741A except that the LM741C/LM741E have their performance guaranteed over a 0 °C to +70 °C temperature range, instead of −55 °C to +125 °C.

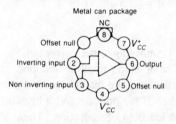

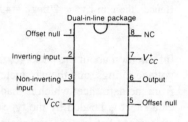

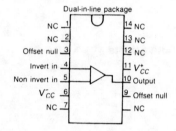

Absolute Maximum Ratings

	LM741A	LM741E	LM741	LM741C
Supply voltage V_{CC}	±22 V	±22 V	±22 V	±18 V
Power dissipation (Note 1)	500 mW	500 mW	500 mW	500 mW
Differential input voltage	±30 V	±30 V	±30 V	±30 V
Input voltage (Note 2)	±15 V	±15 V	±15 V	±15 V
Output short circuit duration	Indefinite	Indefinite	Indefinite	Indefinite
Operating temperature range	−55 °C to +125 °C	0 °C to +70 °C	−55 °C to +125 °C	0 °C to +70 °C
Storage temperature range	−65 °C to +150 °C	−65 °C to +150 °C	−65 °C to +150 °C	−65 °C to +150 °C
Lead temperature (Soldering, 10 seconds)	300 °C	300 °C	300 °C	300 °C

Electrical Characteristics (Note 3)

Parameter	Conditions	LM741A/LM741E			LM741			LM741C			Units
		Min	Typ	Max	Min	Typ	Max	Min	Typ	Max	
Input offset voltage	T_A = 25 °C										
	$R_S \leq 10$ kΩ					1.0	5.0		2.0	6.0	mV
	$R_S \leq 50$ Ω		0.8	3.0							mV
	$T_{Amin} \leq T_A \leq T_{Amax}$										
	$R_S \leq 50$ Ω			4.0							mV
	$R_S \leq 10$ Ω						6.0			7.5	mV
Average input offset voltage drift				15							μV/°C
Input offset voltage adjustment range	T_A = 25 °C, V_{CC} = ±20 V	±10				±15			±15		mV
Input offset current	T_A = 25 °C		3.0	30		20	200		20	200	nA
	$T_{Amin} \leq T_A \leq T_{Amax}$			70		85	500			300	nA
Average input offset current drift				0.5							nA/°C
Input bias current	T_A = 25 °C		30	80		80	500		80	500	nA
	$T_{Amin} \leq T_A \leq T_{Amax}$			0.210			1.5			0.8	μA

Electrical Characteristics (cont.)

Parameter	Conditions	LM741A Min	LM741A Typ	LM741A Max	LM741/LM741E Min	LM741/LM741E Typ	LM741/LM741E Max	LM741C Min	LM741C Typ	LM741C Max	Units
Input resistance	$T_A = 25°C$, $V_{CC} = \pm 20$ V	1.0	6.0		0.3	2.0		0.3	2.0		MΩ
	$T_{Amin} \le T_A \le T_{Amax}$ $V_{CC} = \pm 20$ V	0.5									MΩ
Input voltage range	$T_A = 25$ °C							±12	±13		V
	$T_{Amin} \pm T_A \pm T_{Amax}$				±12	±13					V
Large signal voltage gain	$T_A = 25$ °C, $R_L > 2$ kΩ										
	$V_{CC} = \pm 20$ V, $V_O = \pm 15$ V	50									V/mV
	$V_{CC} = \pm 15$ V, $V_O = \pm 10$ V				50	200		20	200		V/mV
	$T_{Amin} \le T_A \le T_{Amax}$ $R_L \ge 2$ kΩ										
	$V_{CC} = \pm 20$ V, $V_O = \pm 15$ V	32									V/mV
	$V_{CC} = \pm 15$ V, $V_O = \pm 10$ V				25			15			V/mV
	$V_{CC} = \pm 15$ V, $V_O = \pm 12$ V	10									V/mV
Output voltage swing	$V_{CC} = \pm 20$ V										
	$R_L \ge 10$ kΩ	±16									V
	$R_L \ge 2$ kΩ	±15									V
	$V_{CC} = \pm 15$ V										
	$R_L \ge 10$ kΩ				±12	±14		±12	±14		V
	$R_L \ge 2$ kΩ				±10	±13		±10	±13		V
Output short circuit current	$T_A = 25$ °C	10	25	35		25			25		mA
	$T_{Amin} < T_A \le T_{Amax}$	10		40							
Common mode rejection ratio	$T_{Amin} \le T_A \le T_{Amax}$ $R_S \le 10$ kΩ, $V_{CM} = \pm 12$ V				70	90		70	90		dB
	$R_S \le 50$ Ω, $V_{CM} = \pm 12$ V	80	95								dB
Supply voltage rejection ratio	$T_{Amin} \le T_A \le T_{Amax}$ $V_{CC} = \pm 20$ V to $V_{CC} = \pm 5$ V										
	$R_S \le 50$ Ω	86	96								dB
	$R_S \pm 10$ kΩ				77	96		77	96		dB
Transient response	$T_A = 25$ °C, Unit gain										
Rise time		0.25	0.8		0.3			0.3			μs
Overshoot		6.0	20		5			5			%
Bandwidth (Note 4)	$T_A = 25$ °C	0.437	1.5								MHz
Slew rate	$T_A = 25$ °C, Unity gain	0.3	0.7		0.5			0.5			V/μs
Supply current	$T_A = 25$ °C					1.7	2.8		1.7	2.8	mA
Power consumption	$T_A = 25$ °C										
	$V_{CC} = \pm 20$ V	80	150								mW
	$V_{CC} = \pm 15$ V					50	85		50	85	mW
LM741A	$V_{CC} = \pm 20$ V										
	$T_A = T_{Amin}$		165								mW
	$T_A = T_{Amax}$		135								mW
LM741E	$V_{CC} = \pm 20$ V		150								mW
	$T_A = T_{Amin}$		150								mW
	$T_A = T_{Amax}$		150								mW
LM741	$V_{CC} = \pm 15$ V										
	$T_A = T_{Amin}$							60	100		mW
	$T_A = T_{Amax}$							45	75		mW

Note 1: The maximum junction temperature of the LM741/LM741A is 150 °C, while that of the LM741C/LM741E is 100 °C. For operation at elevated temperatures, devices in the TO-5 package must be derated based on a thermal resistance of 150 °C/W junction to ambient, or 45 °C/W junction to case. The thermal resistance of the dual-in-line package is 100 °C/W junction to ambient.

Note 2: For supply voltages less than ± 15 V, the absolute maximum input voltage is equal to the supply voltage.

Note 3: Unless otherwise specified, these specifications apply for $V_{CC} = \pm 15$ V, -55 °C $\le T_A \le +125$ °C (LM741/LM741A). For the LM741C/LM741E, these specifications are limited to 0 °C $\le T_A \le +70$ °C.

Note 4: Calculated value from: BW (MHz) = 0.35/Rise time (μs).

Fig. 4.12 Data sheet for LM 741 op-amp (*courtesy National Semiconductor (UK) Ltd*).

National Semiconductor
LM101A/LM201A/LM301A Operational Amplifier

General Description

The LM101A series are general purpose operational amplifiers which feature improved performance over industry standards like the LM709. Advanced processing techniques make possible an order of magnitude reduction in input currents, and a redesign of the biasing circuitry reduces the temperature drift of input current. Improved specifications include:

- Offset voltage 3 mV maximum over temperature (LM101A/LM201A)

- Input current 100 nA maximum over temperature (LM101A/LM201A)

- Offset current 20 nA maximum over temperature (LM101A/LM201A)

- Guaranteed drift characteristics

- Offsets guaranteed over entire common mode and supply voltage ranges

- Slew rate of 10 V/μs as a summing amplifier

This amplifier offers many features which make its application nearly foolproof: overload protection on the input and output, no latch-up when the common mode range is exceeded, freedom from oscillations and compensation with a single 30 pF capacitor. It has advantages over internally compensated amplifiers in that the frequency compensation can be tailored to the particular application. For example, in low frequency circuits it can be overcompensated for increased stability margin. Or the compensation can be optimized to give more than a factor of ten improvement in high frequency performance for most applications.

In addition, the device provides better accuracy and lower noise in high impedance circuitry. The low input currents also make it particularly well suited for long interval integrators or timers, sample and hold circuits and low frequency waveform generators. Further, replacing circuits where matched transistor pairs buffer the inputs of conventional IC op amps, it can give lower offset voltage and drift at a lower cost.

The LM101A is guaranteed over a temperature range of $-55\,°C$ to $+125\,°C$, the LM201A from $-25\,°C$ to $+85\,°C$, and the LM301A from $0\,°C$ to $70\,°C$.

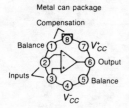

Metal can package

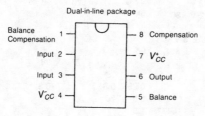

Dual-in-line package

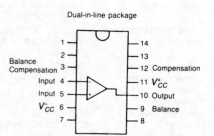

Dual-in-line package

Note: Pin 6 connected to bottom of package.

Absolute Maximum Ratings

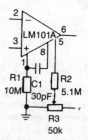

Standard compensation and offset balancing circuit

	LM101A/LM201A	LM301A
Supply voltage V_{CC}	± 22 V	± 18 V
Power dissipation (Note 1)	500 mW	500 mW
Differential input voltage	± 30 V	± 30 V
Input voltage (Note 2)	± 15 V	± 15 V
Output short circuit duration (Note 3)	Indefinite	Indefinite
Operating temperature range	$-55\,°C$ to $+125\,°C$ (LM101A) $-25\,°C$ to $+85\,°C$ (LM201A)	$0\,°C$ to $+70\,°C$
Storage temperature range	$-65\,°C$ to $+150\,°C$	$-65\,°C$ to $+150\,°C$
Lead temperature	$300\,°C$	$300\,°C$

Note 1: The maximum junction temperature of the LM101A is 150 °C, and that of the LM201A/LM301A is 100 °C. For operating at elevated temperatures, devices in the TO-5 package must be derated based on a thermal resistance of 150 °C/W, junction to ambient, or 45 °C/W, junction to case. The thermal resistance of the dual-in-line package is 187 °C/W, junction to ambient.
Note 2: For supply voltages less than ± 15 V, the absolute maximum input voltage is equal to the supply voltage.
Note 3: Continuous short circuit is allowed for case temperatures to 125 °C and ambient temperatures to 75 °C for LM101A/LM201A, and 70 °C and 55 °C respectively for LM301A.
Note 4: Unless otherwise specified, these specifications apply for C1 = 30 pF, ± 5 V $\leq V_{CC} \leq \pm 20$ V and $-55\,°C \leq T_A \leq +125\,°C$ (LM101A), ± 5 V $\leq V_{CC} \leq \pm 20$ V and $-25\,°C \leq T_A \leq +85\,°C$ (LM201A), ± 5 V $\leq V_{CC} \leq \pm 15$ V and $0\,°C \leq T_A \leq +70\,°C$ (LM301A).

Electrical Characteristics (Note 4)

Parameter	Conditions	LM101A/LM201A			LM301A			Units
		Min	Typ	Max	Min	Typ	Max	
Input offset voltage LM101A, LM201A, LM301A	$T_A = 25\ °C$ $R_S \leq 50\ k\Omega$		0.7	2.0		2.0	7.5	mV
Input offset current	$T_A = 25\ °C$		1.5	10		3.0	50	nA
Input bias current	$T_A = 25\ °C$		30	75		70	250	nA
Input resistance	$T_A = 25\ °C$	1.5	4.0		0.5	2.0		MΩ
Supply current	$T_A = 25°C$							
	$V_{CC} = \pm 20\ V$		1.8	3.0				mA
	$V_{CC} = \pm 15\ V$					1.8	3.0	mA
Large signal voltage gain	$T_A = 25°C,\ V_{CC} = \pm 15\ V$ $V_{out} = \pm 10\ V,\ R_L \geq 2\ k\Omega$	50	160		25	160		V/mV
Input offset voltage	$R_S \leq 50\ k\Omega$			3.0			10	mV
	$R_S \leq 10\ k\Omega$							mV
Average temperature coefficient of input offset voltage	$R_S \leq 50\ k\Omega$ $R_S \leq 10\ k\Omega$		3.0	15		6.0	30	μV/°C μV/°C
Input offset current				20			70	nA
	$T_A = T_{max}$							nA
	$T_A = T_{min}$							nA
Average temperature coefficient of input offset current	$25\ °C \leq T_A \leq T_{max}$ $T_{min} \leq T_A \leq 25\ °C$		0.01 0.02	0.1 0.2		0.01 0.02	0.3 0.6	nA/°C nA/°C
Input bias current				0.1			0.3	μA
Supply current	$T_A = T_{max},\ V_{CC} = \pm 20\ V$		1.2	2.5				mA
Large signal voltage gain	$V_{CC} = \pm 15\ V,\ V_{out} = \pm 10\ V$ $R_L \geq 2\ k$	25			15			V/mV
Output voltage swing	$V_{CC} = \pm 15\ V$							
	$R_L = 10\ k\Omega$	±12	±14		±12	±14		V
	$R_L = 2\ k\Omega$	±10	±13		±10	±13		V
Input voltage range	$V_{CC} = \pm 20\ V$	±15						V
	$V_{CC} = \pm 15\ V$		+15	−13	±12	+15	−13	V
Common mode rejection ratio	$R_S \leq 50\ k\Omega$	80	96		70	90		dB
	$R_S \leq 10\ k\Omega$							dB
Supply voltage rejection ratio	$R_S \leq 50\ k\Omega$	80	96		70	96		dB
	$R_S \leq 10\ k\Omega$							dB

Guaranteed Performance Characteristics LM301A

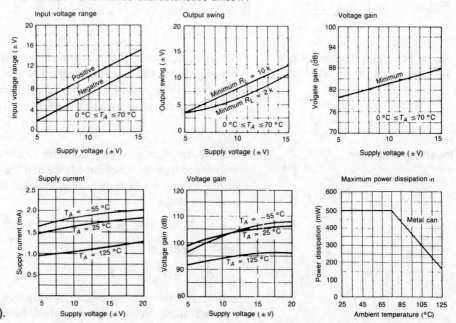

Fig. 4.13 Data sheet for LM 101/201/301 op-amp (*courtesy National Semiconductor (UK) Ltd*).

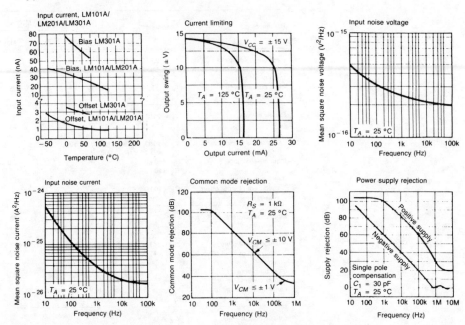

Fig. 4.13 cont.

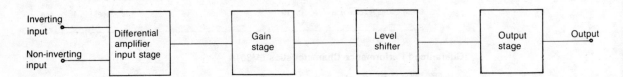

Fig. 4.14 Block diagram of the internal circuitry of an op-amp.

positive or negative when an input signal is applied. Finally, the output stage is usually a complementary emitter-follower stage to provide the op-amp with a low output resistance. It is usually given a small bias current to avoid crossover distortion.

Inverting Amplifier

The circuit of an inverting amplifier is shown in Fig. 4.2(a). If it is assumed that the input impedance and the open-loop voltage gain of the op-amp are both very high and that the output impedance is very low, then

$$V_{in}/R_1 = -V_{out}/R_2 \quad \text{or} \quad A_{v(f)} = V_{out}/V_{in} = -R_2/R_1.$$

Input impedance

$$Z_{in(f)} = V_{in}/I_{in} = R_1.$$

If the first two assumptions are not made, then Fig. 4.15 will show the equivalent circuit of the amplifier. From this

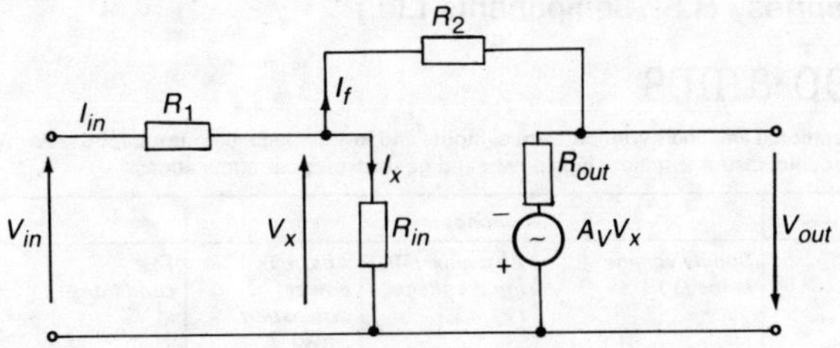

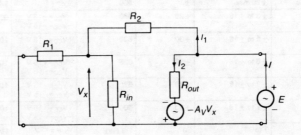

Fig. 4.15 Equivalent circuit of an inverting op-amp.

$$\frac{V_{in}-V_x}{R_1} + \frac{V_{out}-V_x}{R_2} = \frac{V_x}{R_{in}}.$$

But $V_x = -V_{out}/A_v$ so

$$\frac{V_{in}+V_{out}/A_v}{R_1} + \frac{V_{out}(1+1/A_v)}{R_2} = \frac{-V_{out}}{A_v R_{in}}$$

$$V_{in}\left[\frac{1}{R_1}\right] = -V_{out}\left[\frac{1}{A_v R_{in}} + \frac{1}{A_v R_1} + \frac{1+1/A_v}{R_2}\right]$$

$$= -V_{out}\left[\frac{1}{A_v R_{in}} + \frac{1}{A_v R_1} + \frac{1+A_v}{A_v R_2}\right]$$

Therefore,

$$A_{v(f)} = \frac{V_{out}}{V_{in}} = -\frac{1}{R_1} \times \frac{1}{\dfrac{1}{A_v R_{in}} + \dfrac{1}{A_v R_1} + \dfrac{1+A_v}{A_v R_2}}$$

$$= -\frac{R_2}{R_1} \times \frac{1}{\dfrac{1}{A_v}\left[\dfrac{R_2}{R_{in}} + \dfrac{R_2}{R_1} + 1 + A_v\right]}$$

$$A_{v(f)} = -\frac{R_2}{R_1} \times \frac{1}{1+\dfrac{1}{A_v}\left[1+\dfrac{R_2}{R_{in}} + \dfrac{R_2}{R_1}\right]}. \qquad (4.16)$$

The factor $1+\dfrac{1}{A_v}\left[1+\dfrac{R_2}{R_{in}} + \dfrac{R_2}{R_1}\right]$ is the *gain error*.

If A_v is large equation 4.16 reduces to equation (4.1).

If a resistor is connected between the + terminal and earth (Fig.

Fig. 4.16 Calculation of op-amp output resistance.

Table 4.1 (Courtesy R.S. Components Ltd.)

Bi-polar op-amps

A range of bi-polar operational amplifiers with differential inputs and low-impedance output, for use as inverting, non-inverting and differential amplifiers. The range offers amplifiers to suit audio, instrumentation, high power and general-purpose applications.

device	ratings			typical characteristics at 25 °C				
	Supply voltage range (V)	Abs. max. diff. input voltage (V)	Abs. max. power dissipation (mW)	Test conditions	Open-loop voltage gain (dB)	I/P bias current (nA)	Slew rate, response time (V/μs), μs	O/P voltage swing (V)
AD548/648	±4·5 to ±18	—	—	$V_{CC} = \pm 15$ V	100 min	0·01	1·8	±13
L165	±6 to ±18	±15	20 (W)	$V_{CC} = \pm 15$ V	80	200	6	24
L272	4 to 28	±28	1 (W)	$V_{CC} = 24$ V	70	300	1	23
LM11	±2·5 to ±20	1	500	$V_{CC} = \pm 15$ V	108	0·04	0·3	±12
LM301	±5 to ±18	30	500	$V_{CC} = \pm 15$ V	88	70	0·4	±13
LM308	±5 to ±18	30	500	$V_{CC} = \pm 15$ V	102	1·5	—	±13
LM324	3 to 32	32	625	$V_{CC} = +5$ V	100*	45	—	28 or ±14
LM348	±10 to ±18	24	500	$V_{CC} = +5$ V	96*	30	0·6	28 or ±14
LM358	3 to 30	32	570	$V_{CC} = +5$ V	100*	40	0·6	V^+ −1·5 max.
LM725	±4 to ±22	±5	500	$V_{CC} = \pm 15$ V	130	42	0·25	±13·5
LM6361	4·75 to 32	±8	—	$V_{CC} = \pm 15$ V	117	2000	300	$V_O^+ = +14·2$ $V_O^- = -13·4$
LM6364	4·75 to 32	±8	—	$V_{CC} = \pm 15$ V	68	2500	300	$V_O^+ = +14·2$ $V_O^- = -13·4$
LM6365	4·75 to 32	±8	—	$V_{CC} = \pm 15$ V	80	2500	300	$V_O^+ = +14·2$ $V_O^- = -13·4$
LP324	3 to 32	32	—	$V_{CC} = +5$ V	100	2	0·014	3·5
MC33078/33079	+36 V V_{CC} to V_{EE}	30	1·27 W	$V_{CC} = +15$ V $V_{EE} = -15$ V	110	300	7	13·8
MC33171/33172/33174	+3 to +44 ±1·5 to ±22	44	—	$V_{CC} = +15$ v $V_{EE} = -15$ V	114	20	2·1	14·2
NE531	±5 to ±22	15	300	$V_{CC} = \pm 15$ V	96	400	35	±15
NE5532	±3 to ±20	±0·5	1·2 W	$V_{CC} = \pm 15$ V	100	200	9	±13
NE5534	±3 to ±20	±0·5	800	$V_{CC} = \pm 15$ V	100	500	13	±13·5
NE5539	±8 to ±12	—	550	$V_{CC} = \pm 8$ V	52	5000	600	+2·7 to −2·2
OP-07	±3 to ±18	±30	500	$V_{CC} = \pm 15$ V	132	±2·2	0·17	±13
OP-27	±4 to ±18	±0·7	500	$V_{CC} = \pm 15$ V	123	±15	2·8	±13
OP-37	±4 to ±18	±0·7	500	$V_{CC} = \pm 15$ V	123	±15	17	±13
OP-77	±3 to ±18	±30	500	$V_{CC} = \pm 15$ V	135	1·2	0·3	±13
OP-90	+1·6 to +36 ±0·8 to ±18	V^- −20 to V^+ +20	500	$V_{CC} = \pm 1.5$ to ±15	106	4	0·012	±12
OP-97	±2·25 to ±20	±1	500	$V_{CC} = \pm 15$ V	126	±0·03	0·2	±14
OP-200	±20 max.	±30	500	$V_{CC} = \pm 15$ V	130	0·1	0·15	±12·2
OP-400	±20 max.	±30	500	$V_{CC} = \pm 15$ V	130	0·75	0·15	±12·2
OP-470	±18 max.	±1	500	$V_{CC} = \pm 15$ V	119	25	2	±13
OP-471	±18 max.	±1	500	$V_{CC} = \pm 15$ V	108	25	8	±13
OP-490	±18 max.	V^- −20 to V^+ +20	500	$V_{CC} = \pm 1.5$ to ±15	106	4·2	0·012	±11·5
PM-1008	±20 max.	±1	500	$V_{CC} = \pm 15$ V	115	±0·03	0·2	±14
PM-1012	±20 max.	±1	500	$V_{CC} = \pm 15$ V	120	±0·03	0·2	±14
RC4558	±3 to ±18	±30	680	$+V_{CC} = +15$ V $-V_{CC} = -15$ V	85	150	1·7	±13
741CP/N	±5 to ±18	±30	500	$+V_{CC} = +15$ V $-V_{CC} = -15$ V	106	80	0·5	±13
741S	±5 to ±18	±30	625	$+V_{CC} = +15$ V $-V_{CC} = -15$ V	100	200	20	±13
747	±7 to ±18	±30	670	$+V_{CC} = +15$ V $-V_{CC} = -15$ V	106	80	0·5	±13
748	±7 to ±18	±30	500	$+V_{CC} = +15$ V $-V_{CC} = -15$ V	106	80	0·8	±13
μA759	7 to 36 ±3·5 to ±18	30	1·3 (W)	$V_{CC} = \pm 15$ V	106	50	0·5	±12·5

F.E.T. input op-amps

A range of F.E.T. input operational amplifiers offering very low input bias and offset currents.

device	ratings			typical characteristics at 25 °C				
	Supply voltage range (V)	Abs. max. diff. input voltage (V)	Abs. max. power dissipation (mW)	Test conditions	Open-loop voltage gain (dB)	I/P bias current (nA)	Slew rate (V/μs)	O/P voltage swing (V)
AD711/712	±4·5 to ±18	—	—	$V_{CC} = \pm 15$ V	100	0·025	20	±13
CA3130	+6 to +16 ±3 to ±8	±8	630	$V_{CC} = +15$ V	110	0·005	10	13
CA3140	+4 to +36 ±2 to ±18	±8	630	$V_{CC} = +15$ V	100	0·005	9	13
CA3240	+4 to +36 ±2 to ±18	±8	630	$V_{CC} = +15$ V	100	0·005	9	13
LF347/351/353	±5 to ±18	±30	500	$V_{CC} = +15$ V	110	0·05	13	±13·5
LF355	±4 to ±18	±30	500	$V_{CC} = +15$ V	106	0·03	5	±13
LF411	+18 abs. max.	±30	—	$V_{CC} = +15$ V $V_{EE} = -15$ V	98	0·06	25	$V_O^+ = 13·9$ $V_O^- = -14·7$
LF412	+18 abs. max.	±30	—	$V_{CC} = +15$ V $V_{EE} = -15$ V	103	0·05	13	$V_O^+ = 14$ $V_O^- = -14$
OP-42	±20 abs. max.	40	500	$V_{CC} = \pm 15$ V	108	0·13	50	+12·5 −11·9
TL061/062/064	±2 to ±18	±30	680	$V_{CC} = \pm 15$ V	76	0·03	3·5	±13·5
TL071/072/074	±3 to ±18	±30	680	$V_{CC} = \pm 15$ V	106	0·03	13	±13·5
TL081/082/084	±3 to ±18	±30	680	$V_{CC} = \pm 15$ V	106	0·03	13	±13·5
OPA111	±5 to ±18	±36	500	$V_{CC} = \pm 15$ V	125	±0·0008	2	±12
OPA121	±5 to ±18	±36	500	$V_{CC} = \pm 15$ V	114	±0·001	2	±12
OPA606	±5 to ±18	±36	500	$V_{CC} = \pm 15$ V	110	±0·008	30	±12

C-MOS op-amps

A range of CMOS operational amplifiers offering the advantages of very high input resistance and low input currents together with very low supply voltage operation and micropower consumption. Included in the range are BiMOS amplifiers which combine C-MOS and bi-polar technologies, both on a single monolithic chip. In this form, the user benefits from the F.E.T. input stage of the amplifier which provides high input impedance and a wide common-mode input voltage range as well as benefiting from the bi-polar output stage which provides high output current capability.

device	ratings			typical characteristics at 25 °C				
	Supply voltage range (V)	Abs. max. diff. input voltage (V)	Abs. max. power dissipation (mW)	Test conditions	Open-loop voltage gain (dB)	I/P bias current (nA)	Slew rate, response time (V/μs), μs	O/P voltage swing (V)
ICL7611	18 Abs. max.	$\pm[(V^+ +0·3) - (V^- -0·3)]^*$	250	$V_S = \pm 5$ V $R_L = 10$ kΩ	98	0·001	1·6	±4·5
ICL7641	18 Abs. max.	$\pm[(V^+ +0·3) - (V^- -0·3)]^*$	375	$V_S = \pm 5$ V $R_L = 10$ kΩ	98	0·001	1·6	±4·5
ICL7642	18 Abs. max.	$\pm[(V^+ +0·3) - (V^- -0·3)]^*$	375	$V_S = \pm 5$ V $R_L = 1$ MΩ	104	0·001	0.016	±4·9 ($R_L = 100$ kΩ)
ICL7650	±2·5 to ±8	$\pm[(V^+ +0·3) - (V^- -0·3)]^*$	375	$V^+ = +5$ V $V^- = -5$ V	134	0·0015	2·5	±4·85
ICL7652	±2·5 to ±8	$\pm[(V^+ +0·3) - (V^- -0·3)]^*$	375	$V^+ = +5$ V $V^- = -5$ V	150	0·015	0·5	±4·85
TLC251	1 to 16	±18	725	$V_{DD} = 10$ V	109	0·001	0·6	8·6
TLC271	3 to 16	±18	725	$V_{DD} = 10$ V	109	0·001	0·6	8·6
TLC272	3 to 16	±18	725	$V_{DD} = 10$ V	92	0·001	4·5	8·6
TLC274	3 to 16	±18	875	$V_{DD} = 10$ V	92	0·001	4·5	8·6
CA3160	5 to 16 +2·5 to +8	±8	630	$V^+ = 15$ V $V^- = 0$ V	110	0·005	10	$V_O^+ = 13·3$ $V_O^- = 0·002$
CA3240	4 to 36 ±2 to ±18	±8	630	$V^+ = 15$ V $V^- = -15$ V	100	0·01	9	+13 −14·4
CA3260	4 to 16 ±2 to ±8	±8	630	$V^+ = 15$ V $V^- = 0$ V	110	0·005	10*	$V_O^+ = 13·3$ $V_O^- = 0·002$
CA5130	5 to 16 ±2·5 to ±8	±8	630	$V^+ = 15$ V $V^- = 0$ V	110	0·005	10*	$V_O^+ = 13·3$ $V_O^- = 0·002$
CA5160	5 to 16 ±2·5 to ±8	±8	630	$V^+ = 5$ V $V^- = 0$ V	102	0·002	10*	$V_O^+ = 13·3$ $V_O^- = 0·002$
CA5260	4·5 to 16 ±2·25 to ±8	±8	630	$V^+ = 15$ V $V^- = 0$ V	80	0·002	8	$V_O^+ = 4·7$ $V_O^- = 0$ V

* $V^+ = +7·5$ V, $V^- = -7·5$ V

4.4) it is in series with the *much* larger R_{in} and does not affect the gain expression.

Input Resistance

The input resistance $R_{in(f)}$ of an inverting amplifier is the ratio

$$V_{in}/I_{in} = \frac{I_{in}R_1 + V_x}{I_{in}} = R_1 + \frac{V_x}{I_{in}}.$$

Now

$$I_{in} = I_x + I_f$$

and

$$I_f = \frac{V_x(1+A_v)}{R_2 + R_{out}} = \frac{V_x(1+A_v)}{R_2},$$

where $R_2^1 = R_2 + R_{out}$, and $I_x = V_x/R_{in}$.

Therefore,

$$I_{in} = V_x\left(\frac{1+A_v}{R_2^1} + \frac{1}{R_{in}}\right)$$

and

$$R_{in(f)} = R_1 + \left(\frac{R_2^1 R_{in}}{R_{in}(1+A_v) + R_2^1}\right). \tag{4.17}$$

This means that the input impedance of the op-amp itself is reduced by the applied negative feedback. If A_v is large then $R_{in(f)} \simeq R_1$.

Output Resistance

To derive an expression for the output impedance of the inverting amplifier a voltage source of e.m.f. E volts and zero internal resistance is connected across the output terminals of the circuit and the input terminals are shorted together, see Fig. 4.16. The output impedance of the amplifier is then equal to the ratio E/I ohms. From the figure,

$$I_1 = \frac{E}{R_2 + \dfrac{R_1 R_{in}}{R_1 + R_{in}}} = \frac{E}{R_2 + R_T},$$

where

$$R_T = \frac{R_1 R_{in}}{R_1 + R_{in}} \quad \text{and} \quad I_2 = \frac{E + A_v V_x}{R_{out}}, \quad \text{and} \quad V_x = \frac{E R_T}{R_2 + R_T}.$$

Therefore,

$$R_{out(f)} = \cfrac{E}{\cfrac{E}{R_S+R_T} + \cfrac{E+A_vER_T/(R_2+R_T)}{R_{out}}}$$

$$= \cfrac{R_{out}}{\cfrac{R_{out}}{R_S+R_T} + 1 + A_vR_T/(R_2+R_T)}. \tag{4.18}$$

Non-inverting Amplifier

Figure 4.2(*b*) shows an op-amp connected to form *non-inverting amplifier*. Assuming initially that the input impedance of the op-amp is very high and the output impedance is very low, then

$$V_{out} = A_v(V_{in} - V_y)$$

where V_y is the voltage that appears at the negative terminal and A_v is the open-loop voltage gain of the op-amp.

$$V_y = V_{out}R_1/(R_1+R_2). \tag{4.19}$$

Therefore,

$$V_{out} = A_v[V_{in} - V_{out}R_1/(R_1+R_2)]$$

$$V_{out}[1+A_vR_1/(R_1+R_2)] = A_vV_{in}$$

and

$$A_{v(f)} = V_{out}/V_{in} = \frac{A_v}{1+A_vR_1/(R_1+R_2)} \tag{4.20}$$

$$A_{v(f)} = \cfrac{1}{1+\cfrac{1}{A_v}\left(1+R_2/R_1\right)} \frac{R_1+R_2}{R_1}. \tag{4.21}$$

The gain error is $\left[1+\dfrac{1}{A_v}(1+R_2/R_1)\right] + \dfrac{1}{\text{c.m.r.r.}}$.

Input Impedance

To determine an equation for the input impedance of the non-inverting op-amp consider Fig. 4.17

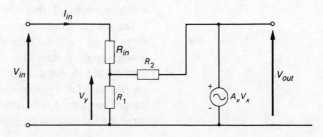

Fig. 4.17 Equivalent circuit of a non-inverting op-amp.

$$R_{in(f)} = V_{in}/I_{in}$$

where $I_{in} = (V_{in} - V_y)/R_{in}$

so $R_{in(f)} = \dfrac{V_{in}R_{in}}{V_{in} - V_y} = \dfrac{R_{in}}{1 - V_y/V_{in}}.$

From equation (4.20)

$$V_{in} = \frac{V_{out}[1 + A_v R_1/(R_1 + R_2)]}{A_v}$$

and from equation (4.19)

$$V_y = V_{out}R_1/(R_1 + R_2).$$

Hence

$$V_y/V_{in} = \frac{R_1 A_v}{(R_1 + R_2)\left[1 + \dfrac{A_v R_1}{R_1 + R_2}\right]} = \frac{A_v R_1}{R_2 + R_1(1 + A_v)}$$

and

$$R_{in(f)} = \frac{R_{in}}{1 - \dfrac{A_v R_1}{R_2 + R_1(1 + A_v)}} \simeq \frac{R_{in}R_2 + R_{in}R_1 A_v}{R_1 + R_2}$$

$$\simeq \frac{R_{in}R_1 A_v}{R_1 + R_2} = \frac{A_v R_{in}}{1 + R_2/R_1}. \tag{4.22}$$

Since $A_v \gg 1 + R_2/R_1$, the input impedance of the non-inverting op-amp is considerably increased. This result should have been expected since it is clear from Fig. 4.17 that voltage-series n.f.b. has been applied to the circuit. The output impedance of the circuit is reduced to a very low value, the analysis being identical to that given for the inverting amplifier.

Single-polarity Power Supply

There are a few op-amps on the market that have been designed to operate from a single-polarity power supply, examples being the LM 10, LM 660 and the CA 3140. All other op-amps are designed to operate from dual $\pm V_{CC}$ power supplies but, if a.c. signals only are to be amplified, they can be worked from a single positive supply.

The output terminal of the op-amp must be held at a convenient value of d.c. voltage which is usually chosen to be one-half of the power supply voltage, i.e. $V_{CC}/2$. To obtain this, the non-inverting terminal is connected to the junction of two resistors R_3 and R_4 which are connected as a potential divider between the positive and the earth lines. The method is illustrated by Figs. 4.18(a) and (b) for both inverting and non-inverting amplifiers. If $R_3 = R_4$ the d.c. potential

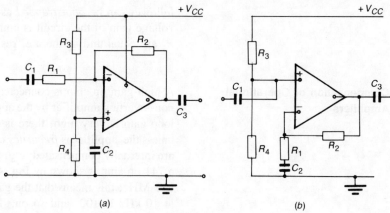

Fig. 4.18 Single-polarity power supply (a) inverting amplifier, (b) non-inverting amplifier.

at the output terminal will be equal to $+V_{CC}/2$. Any a.c. input signal will then vary the output voltage either side of $+V_{CC}/2$, with limits of about $(V_{CC}-1)$ and $+1$ V.

The Voltage Follower

The circuit given in Fig. 4.19(a) is known as a voltage follower. The circuit has 100% n.f.b. applied to it and so it has a voltage gain of unity, a very high input resistance, and a low output resistance. The gain error is $1/A_v$, or $1/c.m.r.r.$, whichever is the smaller. The circuit may need to be compensated for unity-gain operation, see page 122.

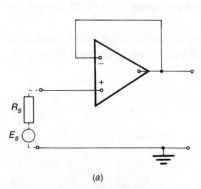

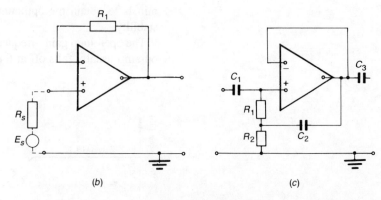

(a)   (b)   (c)

Fig. 4.19 Voltage follower (a) basic circuit, (b) with offset voltage reduction, (c) bootstrapped.

Output offset error caused by the input bias current flowing in the source resistance R_s can be minimized by the connection of a resistor R_1 between the negative and output terminals, as shown in Fig. 4.19(b). R_1 should be of the same resistance value as the source.

If an a.c. voltage follower is used the input may be capacitively coupled. It will then be necessary to provide a d.c. path between the positive terminal and earth. If this is achieved by means of a resistor the input resistance of the circuit will be reduced. To avoid this the follower can be *bootstrapped* as shown by Fig. 4.19(*c*). Since the voltage gain of the circuit is unity the a.c. voltages at either end of R_1 are equal and so the a.c. resistance of R_1 is extremely high.

Compensation of Operational Amplifiers

A high-gain op-amp is connected for operation as an inverting or a non-inverting amplifier by the application of n.f.b. Because its open-loop gain is very high, there is always the risk of gain instabililty unless the amplifier is *frequency compensated*. Many types of op-amp are internally compensated; e.g. the 741. The open-loop response of a 741 op-amp is shown by Fig. 4.19. The gain—bandwidth product is 1 MHz; this means that the gain at 1 MHz is unity, and the gain at 10 kHz is 100, and so on. This is a gain—frequency slope of −20 dB/decade and is produced by a single 30 pF compensating capacitor within the chip.

If n.f.b. is applied to the device to reduce its gain to 60 dB, then the response is flat (because $\beta A_v \gg 1$ and $A_{v(f)} \simeq 1/\beta$) up to the point at which the closed-loop gain line intersects with the open-loop gain curve. It can be seen that this will occur at approximately 1 kHz. Similarly, if the closed-loop gain is set to 40 dB, the gain is flat up to approximately 10 kHz. Note that in each case the gain—bandwidth product is constant. Internally compensated op-amps are of limited bandwidth and slew rate.

Many other op-amps are *not* internally frequency compensated but, instead, are provided with one, or more, compensation terminals which appear on the IC package pins. The designer of an op-amp circuit can then connect external compensation components to these terminals to obtain the optimum combination of both bandwidth *and* stability.

The open-loop gain—frequency characteristic of an uncompensated op-amp usually falls off at 6 dB/octave above the first break point,

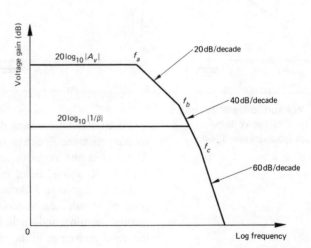

Fig. 4.20 Bode plot of an op-amp open-loop gain response.

and then falls at 12 dB/octave; and, in some cases, also rolls off at 18 dB/octave at frequencies higher than a third break point. Figure 4.20 shows a Bode plot of the open-loop response of such an op-amp; clearly the amplifier has three break frequencies, labelled respectively as f_a, f_b and f_c. When n.f.b. is applied, the closed-loop gain is approximately equal to $1/\beta$ as long as $\beta A_v \gg 1$ and this curve is also plotted. At the intersection of the two curves

$$20\log_{10}|A_v| = 20\log_{10}|1/\beta| \qquad 20\log_{10}|\beta A_v| = 0$$

or $|\beta A_v| = 1$.

For stability the difference in slope between the two curves at the point of intersection must be *less than* 12 dB/octave or 40 dB/decade.

If only a small amount of feedback is applied, the point of intersection will occur where the open-loop gain has a slope of only −6 dB/octave or 20 dB/decade. The circuit is then inherently stable and so no compensation is needed.

The basic requirement for successful frequency compensation is to ensure that the fall-off in gain is *less than* 12 dB/octave when the loop gain βA_v is unity. For *unconditional stability* the gain–frequency characteristic should fall at 6 dB/octave until it reaches the point of unity gain.

(*a*) Suppose a capacitor (usually some 5−50 pF) is connected across the appropriate terminals to introduce a break frequency f_1 which is lower than the lowest uncompensated break frequency f_a. The effect of this, shown by Fig. 4.21, is to ensure that the point of intersection occurs where the open-loop gain has a slope of only −6 dB/octave. The capacitance value must be chosen so that the new frequency of intersection is lower than f_b. This method of compensation reduces both the bandwidth and the slew rate of the amplifier.

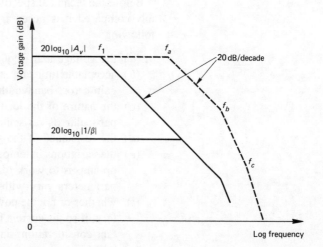

Fig. 4.21 Frequency compensation of an op-amp.

(*b*)Another method of compensation consists of shifting the second break frequency f_b to a new frequency f_1 that is lower than f_a by means of an *RC* series circuit connected across the appropriate pins. The capacitor C causes the gain to fall off at 6 dB/octave at f_1 up to f_a; above f_a the resistance R predominates and nullifies the effect of C. This means that at frequencies higher than f_a the characteristic is the inherent response of the op-amp.

(*c*)One cause of decreasing op-amp gain with increase in frequency is stray capacitance at the input terminals. The effect of this can be compensated for by the connection of a capacitor in parallel with the feedback resistor.

When applying frequency compensation to an op-amp the best procedure is to refer to the data sheet for the device, and to follow its recommendations. The data sheet for the 101/201/301 gives three different methods of compensation together with graphs that show their effects upon the gain of the op-amp; these are shown in Fig. 4.22. A greater stability can always be obtained by the use of a capacitor of larger value than that recommended, at the expense of a reduction in bandwidth, but a smaller value ought never to be used. Single-pole compensation gives a unity-gain bandwidth of 1 MHz with $C = 30$ pF and of 10 MHz with $C = 3$ pF; the latter should not be used if the closed-loop gain is less than 20 db. Two-pole compensation gives a 12 dB/octave fall-off in gain.

Selection of an Op-amp

There is a very large number of different op-amps that are available from various sources, and the choice of a device for a particular application should take into account (*a*) the technical specification, (*b*) the cost, (*c*) the availability, and (*d*) whether or not the op-amp is second sourced. The technical information on an op-amp can readily be obtained from distributor's concise data (e.g. Table 4.1), and then from the data sheets of specific ICs.

Before the technical specification of various op-amps can be sensibly compared it is necessary to have a clear idea of each of the following:

(*a*) the voltage and frequency ranges of the input signal;
(*b*) acceptable limits for such parameters as output offset voltage, gain error, bandwidth and slew rate;
(*c*) the nature of the load into which the op-amp is to work, in particular its resistance and its capacitance;
(*d*) the resistance of the signal source;
(*e*) the variations in temperature of the environment in which the op-amp is to work (data sheets often show how the various parameters vary with change in temperature);
(*f*) whether or not the power supplies will be derived from a battery, if so the current taken from the supply will be an important consideration; and
(*g*) whether or not low noise is of paramount importance.

(a) Single-pole compensation

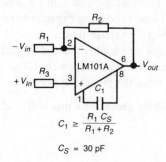

$$C_1 \geq \frac{R_1 \, C_S}{R_1 + R_2}$$

$$C_S = 30 \text{ pF}$$

(b) Two-pole compensation

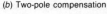
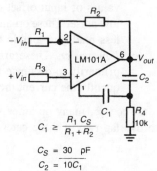

$$C_1 \geq \frac{R_1 \, C_S}{R_1 + R_2}$$

$$C_S = 30 \text{ pF}$$
$$C_2 = \frac{C_1}{10}$$

(c) Feedforward compensation

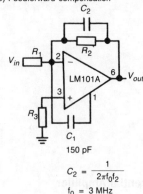

$$C_1$$
$$150 \text{ pF}$$

$$C_2 = \frac{1}{2\pi f_0 f_2}$$

$$f_0 = 3 \text{ MHz}$$

**Pin connections shown are for metal can.

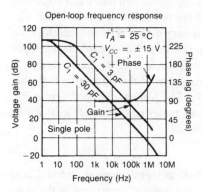

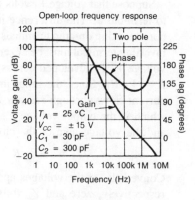

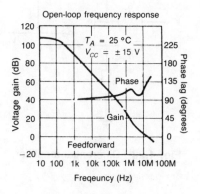

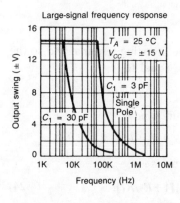

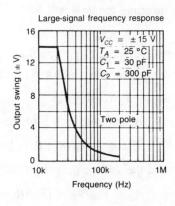

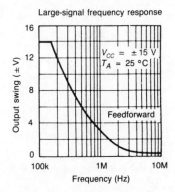

Fig. 4.22 Frequency compensation for the LM 101/201/301 op-amps (a) single-pole, (b) two-pole, (c) feedforward. (*Courtesy National Semiconductor (UK) Ltd*).

There are three basic categories into which op-amps may be placed. These are shown below.

(a) *General purpose* (e.g. 741, 101/201/301). These are the cheapest and most readily available but do not have their performance optimized in any particular respect. A general-purpose op-amp should be selected whenever a high gain, or very low offset, or high speed are not essential.

(b) *High accuracy*. This category refers to op-amps with low values of input offset current and voltage, low c.m.r.r. and low noise. These op-amps will tend to be more expensive and less readily available than general-purpose types.

(c) *Low power*. These are op-amps which have been designed primarily to dissipate the minimum possible power. Consequently the current they take from the power supply is low.

Of course, many op-amps will possess parameters that will tend to overlap two of these categories.

Differential Amplifier

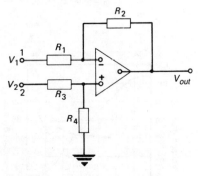

Fig. 4.23 Differential amplifier.

When an op-amp is used as a differential amplifier (Fig. 4.23), voltages are applied to its two input terminals, 1 and 2, and the *difference* between these voltages is amplified.

Suppose that voltage V_1 volts is applied to terminal 1 and zero volts to terminal 2. The difference in the potentials at the inverting and non-inverting op-amp terminals is very nearly zero and therefore the inverting terminal must be at zero potential. This means that the input voltage V_1 is developed across resistor R_1 and the input current is $I_1 = V_1/R_1$. Since the input impedance of the op-amp is high this current flows through resistor R_2. The voltage dropped across R_2, which is the output voltage V_{out} of the circuit, is equal to $V_1 R_2/R_1$ and the voltage gain of the circuit is

$$A_v = V_{out}/V_1 = -R_2/R_1. \qquad (4.23)$$

Conversely, if the voltages applied to input terminals 1 and 2 are, respectively, zero and V_2 volts, the voltage appearing at the non-inverting terminal will be

$$V_2 R_4/(R_3+R_4) \text{ volts.}$$

This voltage will also appear at the inverting terminal and so the voltage across resistor R_1 must be equal to

$$-V_2 R_4/(R_3+R_4).$$

The output voltage V_{out} of the circuit is now

$$V_{out} = V_2 R_4/(R_3+R_4)+[-V_2 R_4/(R_3+R_4)] \times -R_2/R_1$$

and the voltage gain A_v of the circuit is

$$A_v = V_{out}/V_2 = [R_4/(R_3+R_4)][1+R_2/R_1]. \qquad (4.24)$$

The output voltage when both inputs V_1 and V_2 are present is, combining equations (4.23) and (4.24), given by

$$V_{out} = \frac{R_4}{R_3+R_4}\left(1+\frac{R_2}{R_1}\right)V_2 - \frac{R_2}{R_1}V_1. \qquad (4.25)$$

If $R_1 = R_3$ and $R_2 = R_4$ equation (4.25) reduces to

$$V_{out} = (V_2-V_1)R_2/R_1. \qquad (4.26)$$

The basic differential amplifier suffers from two main disadvantages: (a) it has a fairly low input resistance which is different at each input, and (b) the c.m.r.r. is reduced by any resistor mismatches. This latter statement is easily proved. For a common-mode input signal with $V = V_1 = V_2$ equation (4.25) gives

$$\frac{V_{out}}{V} = \text{common-mode gain} = \frac{R_4}{R_3+R_4}\left(1+\frac{R_2}{R_1}\right)-\frac{R_2}{R_1}.$$

If $R_3 = R_1$ and $R_2 = R_4$, the common-mode gain would be zero. Suppose that $R_2 = R_4 = 20R_1$, but that $R_3 = 0.9R_1$, then

$$\text{common-mode gain} = \frac{20R_1}{0.9R_1+20R_1}\left(1+\frac{20R_1}{R_1}\right)-\frac{20R_1}{R_1}$$

$$= 0.0957$$

and so

$$\text{c.m.r.r.} = 20\,\log_{10}\left(\frac{20}{0.0957}\right) = 46\text{ dB}.$$

Instrumentation Amplifier

An instrumentation amplifier is a development of the basic differential amplifier that is widely employed for the precise measurement of differential signals with a very high c.m.r.r. The amplifier possesses the valuable property of being able to amplify low-level signals without error in the presence of considerable common-mode noise voltages. It is widely used in measurement systems and for amplifying the output of transducers such as the thermocouple and the strain gauge.

Figure 4.24 shows the circuit of an instrumentation amplifier. A differential amplifier of the type shown in Fig. 4.23 is preceded by two op-amps and resistors R_1, R_2 and R_3. The resistors R_1 and R_3

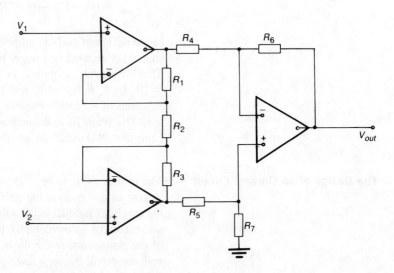

Fig. 4.24

apply voltage feedback to the input on-amps and this reduces their output impedances to a low value, and increases their input impedances to a very high figure. Since the input terminals of each of the input op-amps are at very nearly the same potential the voltage across R_2 is $V_1 - V_2$, and so the current flowing in R_2 is $I = (V_1 - V_2)/R_2$ so that the input voltages to the output op-amp are, respectively,

$$V_1^1 = V_1 + IR_1$$

and

$$V_2^1 = V_2 - IR_3.$$

Hence,

$$V_1^1 = V_1 + \frac{(V_1 - V_2)R_1}{R_2} = V_1\left(1 + \frac{R_1}{R_2}\right) - \frac{V_2 R_1}{R_2}$$

and

$$V_2^1 = V_2 - \frac{(V_1 - V_2)R_3}{R_2} = V_2\left(1 + \frac{R_3}{R_2}\right) - \frac{V_1 R_3}{R_2}.$$

The differential voltage applied to the output op-amp is

$$V_2^1 - V_1^1 = -V_1\left[\left(1 + \frac{R_1}{R_2}\right) - \frac{V_2 R_1}{R_2}\right] + V_2\left[\left(1 + \frac{R_3}{R_2}\right) - \frac{V_1 R_3}{R_2}\right]$$

$$= -V_1\left[1 + \frac{R_1}{R_2} + \frac{R_3}{R_2}\right] + V_2\left[1 + \frac{R_3}{R_2} + \frac{R_1}{R_2}\right].$$

Usually, $R_1 = R_3$ and then

$$V_2^1 - V_1^1 = (V_2 - V_1)(1 + 2R_1/R_2)$$

and if $R_4 = R_5$, $R_6 = R_7$, then

$$V_{out} = (V_2 - V_1)\frac{R_6}{R_4}(1 + 2R_1/R_2). \tag{4.27}$$

In the design of such an amplifier the required voltage gain must be shared between the two stages and this may as well be an equal share. Thus, if the overall gain is to be 100, each stage could have a gain of 10. Then, $R_6/R_7 = 10$ and $1 + 2R_1/R_2 = 10$; there are many combinations of the resistor values that would be suitable. Alternatively, one of the many IC instrumentation amplifiers could be used; examples being the PMI AMP 01 or 05, or the AD 521.

The Design of an Op-amp Circuit

The first decision to be made is whether the circuit is to give an inverting or a non-inverting gain. The input impedance of the inverting amplifier is equal to the value of the input resistor R and this may sometimes be inconveniently low. Conversely, the input resistance of the non-inverting circuit is very high. In either case the output resistance will be very low.

D.C. Amplifier

Inverting Amplifier

Suppose that the wanted output voltage is V_{OUT} volts when the input voltage is V_{IN} volts. The ratio V_{OUT}/V_{IN} will determine the required gain of the circuit and so the ratio of the resistors R_2 and R_1. To minimize the effects of the input bias currents the input current I_{IN} should be made several n times larger than the maximum bias current $I_{B(max)}$ quoted in the data sheet. Then $R_1 = V_{IN}/n I_{B(max)}$. A useful guide is to make $R_1 I_{B(max)} \leq V_{IN}/10$. If the value for R_1 thus obtained is too low for the required minimum input resistance it can be increased, at the expense of increased offset error. If need be another type of op-amp may have to be used. Once the value for R_1 has been settled, then R_2 can be determined. The product $I_{OS}R_2$ should be less than $V_{IN}/10$. It will then be possible to calculate the required value for the resistor R_3 that will further decrease the offset due to the bias currents; $R_3 = R_1 R_2/(R_1 + R_2)$. The feedback resistor is effectively in parallel with the output resistance R_{out} of the op-amp and so it should not be of too low a value, say a minimum of ten times R_{out}, and the op-amp must have the capability to supply a current of at least V_{OUT}/R_2.

Non-inverting Amplifier

The current I that flows in the feedback path $R_1 + R_2$ should be n times larger than the maximum input bias current $I_{B(max)}$. Also $V_{IN}/nI = R_1 + R_2$. The effect of input bias currents should be reduced by the insertion of resistor R_3 in series with the non-inverting terminal with a value such that $R_3 + R_s = R_1 R_2/R_1 + R_2$. For both circuits the input offset voltage V_{OS} is minimized by the recommended nulling circuit given in the data sheet, and frequency compensation may be necessary.

A.C. Amplifier

When the amplifier is to handle a.c. signals only the output offset voltage can be removed by the use of coupling capacitors. It should be checked, though, whether nulling for I_B and V_{OS} is still required. Now the important parameters are the unity-gain bandwidth and the full-voltage power bandwidth, depending upon the magnitude of the anticipated output voltage. The bandwidth will be reduced by the applied n.f.b. and by the coupling capacitors.

Example 4.5

Design an inverting amplifier, using the 301 op-amp, to give an output voltage of ± 2 V when the input voltage is ± 100 mV over the frequency range of 0 to 100 Hz.

Solution

The 301 has a maximum bias current $I_{B(max)}$ of 250 nA. If n is chosen to be 10 the minimum input current I_{IN} will be 2.5 μA and $R_1 = 100 \times 10^{-3}/2.5 \times 10^{-6} = 40$ kΩ. Then $I_{B(max)}R_1 = 0.01$ and $V_{IN}/10 = 0.01$, which just satisfies the quoted design guide. The gain is to be $2/0.1 = 20$ so that $R_2 = 800$ kΩ. Such a large value may lead to noise problems so try again with $n = 100$. Now $I_{IN} = 25$ μA and $R_1 = 4$ kΩ, and $I_{B(max)}R_1 = 0.001$, which is smaller than $V_{IN}/10$ by a factor of 10. Choosing this value for R_1 gives $R_2 = 80$ kΩ, and $R_3 = 3.8$ kΩ. The other design guide is $I_{OS}R_2 \leq V_{IN}/10$; here $50 \times 10^{-9} \times 80 \times 10^3 = 4 \times 10^{-3}$, so this is also satisfied.

The closed-loop gain is 26 dB and Fig. 4.22 shows that single-pole compensation will be satisfactory with $C = 30$ pF.

5 Audio-frequency Large-signal Amplifiers

In an audio-frequency large-signal (power) amplifier the main considerations are the output power, the efficiency, and the percentage distortion. When an appreciable output power is required, a large amplitude input signal is necessary in order to obtain large swings of output current and voltage. The transistors used in a discrete-component circuit must be selected and biased so that their maximum current, voltage, and power ratings are not exceeded when the maximum output power is developed.

For the maximum power to be delivered to a load, without exceeding a predetermined distortion level, the output transistors must work into a particular value of load impedance, known as the optimum load. Power amplifiers using discrete components may employ either bipolar transistors or power FETs and usually operate as a push-pull circuit. In all cases the operation must be restricted to the safe operating area (p. 10). A variety of integrated-circuit power amplifiers are also available, some capable of delivering several watts output power.

Single-ended Power Amplifiers

The output transistor of a single-ended power amplifier should work into its optimum load impedance and this usually demands the use of an output transformer as shown in Fig. 5.1. The turns ratio of the output transformer should be equal to $\sqrt{}$[Optimum load impedance/Actual load impedance].

Bias and d.c. stabilization of a transistor power amplifier is best achieved by means of the potential divider bias circuit, but to minimize d.c. power losses the emitter resistor should be of very low value, perhaps as small as 1 Ω. When such a low value of emitter resistance is used, the resistor is not decoupled because the required capacitance value would be very high. In high-power circuits where the emitter current may be several amperes, the emitter resistor may even be omitted.

Fundamentally, a power amplifier is a converter of d.c. power taken from the power supply into a.c. power delivered to the load. Usually, the amplifier parameter of the greatest importance is its efficiency; the *collector efficiency* η is given by

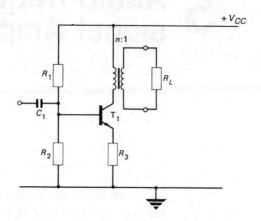

Fig. 5.1 Single-ended power amplifier.

$$\eta = \frac{\text{a.c. power output to load}}{\text{d.c. power taken from power supply}} \times 100\%. \quad (5.1)$$

The d.c. power P_{DC} taken from the power supply is equal to the product of the collector supply voltage and the d.c. component of the collector current, i.e.

$$P_{DC} = V_{CC}I_{C(dc)}. \quad (5.2)$$

The d.c. power provides the a.c. power output plus various power losses within the amplifier itself. Therefore,

$$P_{DC} = P_{out(ac)} + \text{d.c. power losses}. \quad (5.3)$$

Power is lost within the amplifier because of the resistance of the primary winding of the output transformer, the emitter resistor, and dissipation at the collector of the transistor. Very often these losses, except for the last, are small enough to be neglected, and then

$$P_{DC} = P_{out(ac)} + P_c \quad (5.4a)$$

where P_C is the collector dissipation.

Rearranging, $P_c = P_{DC} - P_{out(ac)}.$ $\quad (5.4b)$

The d.c. power taken from the power supply is constant (but see p. 133) and hence the collector dissipation attains its maximum value when the input signal is zero and there is no output power. Care must be taken to ensure that the maximum collector dissipation specified by the manufacturer is not exceeded.

The output power and efficiency of an a.f. power amplifier can be determined with the aid of an a.c. load line drawn on the output characteristics. Since the collector dissipation is at its maximum value under quiescent conditions, the operating point must be chosen to lie on or under the maximum collector dissipation hyperbola.

When a signal is applied to the amplifier, the swings of collector current and voltage are

$$I_{c(max)} - I_{c(min)} \quad \text{and} \quad V_{ce(max)} - V_{ce(min)}$$

so that the a.c. power output is

$$P_{out(ac)} = \tfrac{1}{8}[I_{c(max)} - I_{c(min)}][V_{ce(max)} - V_{ce(min)}] \text{ W}. \tag{5.5}$$

Maximum Collector Efficiency

The collector efficiency η of a power amplifier can be written as

$$\eta = \frac{[I_{c(max)} - I_{c(min)}][V_{ce(max)} - V_{ce(min)}]}{8V_{CC}I_{C(dc)}} \times 100\%. \tag{5.6}$$

The maximum peak a.c. component of the collector current is equal to $I_{C(dc)}$ and the maximum peak value of V_{ce} is equal to V_{CC}. Then

$$I_{c(max)} = 2I_{C(dc)} \quad I_{c(min)} = 0 \quad V_{ce(max)} = 2V_{CC} \quad V_{ce(min)} = 0.$$

Therefore, the maximum value η_{max} of the collector efficiency is

$$\eta_{max} = \frac{2I_{C(dc)} \times 2V_{CC}}{8V_{CC}I_{C(dc)}} \times 100\% = 50\%. \tag{5.7}$$

In practice, V_{ce} cannot be driven down to 0 V nor can I_c very nearly approach either $2I_{C(dc)}$ or 0 without considerable distortion being introduced. For this reason practical efficiencies are always considerably less than the theoretical maximum figure. Typical efficiencies are in the region of 40%

Harmonic Distortion

The mutual characteristic of a large-signal transistor can be represented by a power series

$$I_c = I_{dc} + aV_{be} + bV_{be}^2 + cV_{be}^3 + \text{etc.}, \tag{5.8}$$

where a, b and c are constants and V_{be} is the voltage applied between the base and the emitter. If $V_{be} = V_s \sin \omega t$, and neglecting cubic and higher-order terms for simplicity, then

$$I_c = I_{dc} + aV_s \sin \omega t + bV_s^2 \sin^2 \omega t$$

$$= I_{dc} + \frac{bV_s^2}{2} + aV_s \sin \omega t - \frac{bV_s^2}{2} \cos 2\omega t.$$

This simple analysis shows that the application of a sinusoidal signal to a parabolic transistor mutual characteristic results in both an increase in the d.c. collector current and in the presence of a second-harmonic component. The percentage second-harmonic distortion is the ratio of the amplitudes of the second and fundamental components times 100, i.e. $D_2 = 100I_2/I_f\%$. If the cubic and higher terms are included third-, and higher-order, harmonics will also appear in the analysis. The percentage distortion produced by each harmonic is defined in the same way as for the second harmonic, i.e. the third-harmonic distortion is $D_3 = 100I_3/I_f\%$.

The total harmonic distortion (t.h.d.) is the square root of the sum of the squares of the individual harmonic distortions, i.e.

$$\text{t.h.d.} = \sqrt{(D_2^2 + D_3^2 + D_4^2 + \dots)} \tag{5.9}$$

Class B Push-pull Amplifiers

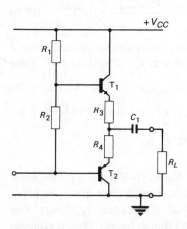

Fig. 5.2 Basic complementary-pair Class B push-pull amplifier.

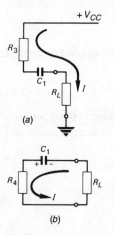

Fig. 5.3

The majority of Class B push-pull amplifiers are of the *complementary-pair type*, shown in Fig. 5.2. Two transistors, one p-n-p and the other n-p-n, are slightly forward-biased and operated as a pair of emitter followers. The quiescent condition of the circuit is with the junction of the emitter resistors at a potential of $V_{CC}/2$.

When an input signal is applied to the circuit, its positive half cycles drive T_1 into conduction and turn T_2 OFF. If the amplitude of the input signal is large enough, T_1 is driven into saturation at the peak of the half cycle. The ON resistance of T_1 is very small and the OFF resistance of T_2 is very high so that the circuit can be redrawn as shown in Fig. 5.3(a). Current flows via R_3 and R_L to charge C_1 to V_{CC} volts and then the voltage across the load R_L is V_{CC} volts. During the negative half cycles of the input signal voltage, T_1 is turned OFF and T_2 is conducting and the circuit can be represented by Fig. 5.3(b). Capacitor C_1 now provides the power supply voltage to T_2. C_1 discharges via R_4 and R_L and, when C_1 has completely discharged, the load voltage is zero.

The collector current of each transistor therefore flows in a series of half-sinewave pulses and the two currents combine at the output to produce a sinusoidal output waveform. The mutual characteristics of a transistor are non-linear for small values of collector current (Fig. 5.4(a)) and this gives rise to *crossover distortion* (see Fig. 5.4(b)). Crossover distortion can be reduced by biasing both transistors to pass a small quiescent collector current. The load voltage varies about its mean value of $V_{CC}/2$, reaching a maximum positive value of V_{CC} volts and a minimum value of zero. The peak value of the a.c. component of the load voltage is $V_{CC}/2$ and the maximum output power is $V_{CC}^2/8R_L$ watts.

If dual power supplies $\pm V_{CC}$ are used the capacitor C_1 is not necessary. The value of C_1 is set by the wanted lower 3 dB frequency f_1 of the circuit and is calculated from $C_1 = 1/(2\pi f_1 R_L)$. The emitter resistors provide thermal stability for the circuit.

The collector current of each transistor flows in a series of half-sinewave pulses and the Fourier analysis of such a waveform shows that its instantaneous value is given by

$$i_c = \frac{I_{c(max)}}{\pi} + \frac{I_{c(max)}}{2}\sin\omega t - \frac{2I_{c(max)}}{3\pi}\cos 2\omega t + \dots \tag{5.10}$$

where $I_{c(max)}$ is the maximum value of the collector current and ω

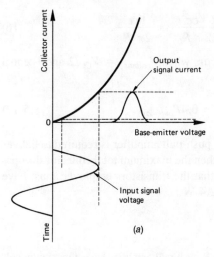

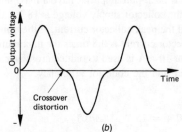

Fig. 5.4 Crossover distortion.

is 2π times the input signal frequency. The first term represents a d.c. component and the second term is the required fundamental frequency component. The fundamental-frequency a.c. power delivered by each transistor to the load is

$$\left(\frac{I_{c(max)}}{2\sqrt{2}}\right)^2 R_L$$

and so the *total* output power at the fundamental frequency is

$$P_{out} = 0.25 I_{c(max)}^2 R_L. \tag{5.11}$$

The d.c. power taken by each transistor from the power supply is $I_{c(max)} V_{CC}/2\pi$ and so the *total* d.c. input power is

$$P_{dc} = I_{c(max)} V_{CC}/\pi. \tag{5.12}$$

The *collector efficiency* η is $100 P_{out}/P_{DC}$. Alternatively, the signal frequency output power can be written as

$$P_{out} = \frac{I_{c(max)}}{2\sqrt{2}} \times \frac{V_{ce(max)}}{\sqrt{2}} \times 2 = \frac{I_{c(max)} V_{ce(max)}}{2}$$

$$= \tfrac{1}{2} I_{c(max)} \left[\frac{V_{CC}}{2} - V_{ce(min)} \right]. \tag{5.13}$$

The collector efficiency η is

$$\eta = 100 P_{out}/P_{DC} = \frac{I_{c(max)} \left[\dfrac{V_{CC}}{2} - V_{ce(min)} \right]}{4 I_{c(max)} V_{CC}/2\pi} \times 100$$

or $\quad \eta = \dfrac{\pi}{4} \left[1 - \dfrac{V_{ce(min)}}{V_{CC}/2} \right] \times 100\%. \tag{5.14}$

Maximum efficiency occurs when the minimum collector−emitter voltage is zero. Then $\eta_{max} = 78.5\%$ but practical efficiencies are some $50{-}60\%$. The total collector dissipation is

$$P_c = P_{DC} - P_{out} = \frac{I_{c(max)} V_{CC}}{\pi} - \frac{I_{c(max)} V_{ce(max)}}{2}$$

$$= \frac{V_{CC}}{\pi} \cdot \frac{V_{ce(max)}}{R_L} - \frac{V_{ce(max)}^2}{2R_L}. \tag{5.15}$$

To determine the maximum possible collector dissipation, differentiate P_c with respect to $V_{ce(max)}$ and equate the result to zero, i.e.

$$\frac{\mathrm{d}P_c}{\mathrm{d}V_{ce(max)}} = \frac{V_{CC}}{\pi R_L} - \frac{V_{ce(max)}}{R_L} = 0.$$

Therefore, $V_{CC}/\pi = V_{ce(max)}$.
Substituting this value of $V_{ce(max)}$ into equation (5.15) gives

$$P_{c(max)} = \frac{V_{CC}^2}{\pi^2 R_L} - \frac{V_{CC}^2}{2\pi^2 R_L} = \frac{V_{CC}^2}{2\pi^2 R_L}. \tag{5.16}$$

Maximum output power occurs when $V_{ce(max)} = V_{CC}/2$ and is equal to $V_{CC}^2/8R_L$ and therefore

$$P_{c(max)} = \frac{4P_{out(max)}}{\pi^2} \simeq 0.4P_{out(max)}. \tag{5.17}$$

If, for example, a Class B push-pull amplifier is required to deliver a power of 12 W to a load, then the maximum total collector dissipation is 4.8 W. This means that the transistors employed must have a power rating of at least 2.4 W.

Example 5.1

The transistors used in a Class B push-pull amplifier have a maximum collector dissipation of 3.0 W. If the collector supply voltage is 18 V calculate the maximum power output and the peak collector current. If the transistors are driven so that the peak collector current is 0.8 times the maximum peak value and a quiescent current of 10 mA is used to reduce crossover distortion, calculate the output power and the collector efficiency.

Solution
Maximum output power $= 6/0.4 = 15$ W.   (*Ans.*)

Therefore $\dfrac{I_{c(max)}}{2} \cdot \dfrac{V_{CC}}{2} = 15$

$$I_{c(max)} = \frac{15 \times 4}{18} = 3.33 \text{ A}. \quad (Ans.)$$

The actual peak collector current is $0.8 \times 3.33 = 2.67$ A and

$$P_{out} = \frac{2.67}{2\sqrt{2}} \times \frac{0.8 \times 9}{\sqrt{2}} \times 2 = 9.61 \text{ W}. \quad (Ans.)$$

The mean collector current per transistor is $2.67/\pi$ A and so the total d.c. power taken from the supply is

$$P_{DC} = \frac{2.67 \times 18}{\pi} + 10 \times 10^{-3} \times 18 = 15.48 \text{ W}.$$

Therefore, the collector efficiency η is

$$\eta = \frac{9.61}{15.48} \times 100 = 62.1\%. \quad (Ans.)$$

The correct operation of the complementary-pair circuit depends upon the quiescent potential at the junction of R_3 and R_4 being held constant at $V_{CC}/2$ volts. This requirement is usually satisfied by the use of d.c. negative feedback from the junction of the emitter resistors to the base of the driver transistor (see Fig. 5.5).

The base bias voltage for T_1 is obtained from the potential divider R_5 and R_6 connected between the output-stage mid-point and earth.

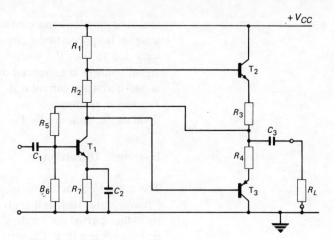

Fig. 5.5 Complementary-pair Class B push-pull amplifier.

If the d.c. voltage at the mid-point should rise, the base bias voltage of T_1 will also rise and T_1 will conduct a larger current. This will make the voltages dropped across R_1 and R_2 increase, making the base potentials of T_2 and T_3 less positive. This, in turn, increases the resistance of T_2 and decreases the resistance of T_3 so that the voltage across T_2 rises while the voltage across T_3 falls. This action will tend to make the mid-point voltage move towards $V_{CC}/2$ volts. The action of the d.c. feedback loop is equally effective in counteracting a fall in the mid-point voltage and so its effect is to stabilize the voltage at the desired value of $V_{CC}/2$ volts.

The d.c. component of the collector current of T_1 must be greater than the peak base current taken by T_2 and T_3. Because of this the maximum possible values of R_1 and R_2 are limited and this sets a limit to the gain of the driver stage. A considerable increase in the gain can be obtained if the stage is *bootstrapped*, as shown by Fig. 5.6. The resistor R_1 has been replaced by two resistors R_{1a} and R_{1b}

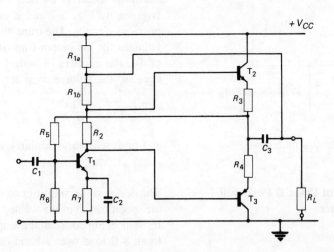

Fig. 5.6 The bootstrapped Class B push-pull amplifier.

and their junction connected to the top of the load resistor R_L. When a signal is applied to the circuit, the emitter potentials of T_2 and T_3 vary and so does the junction of R_{1a} and R_{1b}. This means that the signal voltages at either end of R_{1b} are very nearly equal. Hence the signal-frequency current that flows in R_{1b} is very small and so its effective a.c. resistance is very high. Since the a.c. voltage gain depends upon its a.c. collector load impedance, a large gain is made possible.

Harmonic Distortion

The mutual characteristic of each transistor can be represented by a power series such as that given by equation (5.8). A similar analysis to that carried out for the Class A circuit is valid but there is now no bias voltage (true Class B) and only one transistor conducts at a time. Hence, considerable third-harmonic content exists, and

$$I_{c1} = aV\sin\omega t + bV^2\sin^2\omega t + cV^3\sin^3\omega t$$
$$= aV\sin\omega t + \tfrac{1}{2}bV^2 - \tfrac{1}{2}bV^2\cos 2\omega t + \tfrac{1}{4}3cV^3\sin\omega t - \tfrac{1}{4}cV^3\sin 3\omega t$$
$$= \tfrac{1}{2}bV^2 + (aV + \tfrac{1}{4}3cV^3)\sin\omega t - \tfrac{1}{2}bV^2\cos 2\omega t - \tfrac{1}{4}cV^3\sin 3\omega t.$$

During the alternative half cycles of the input signal,

$$I_{c2} = -aV\sin\omega t + bV^2\sin^2\omega t - cV^3\sin^3\omega t$$
$$= \tfrac{1}{2}bV^2 - (aV + \tfrac{1}{4}3cV^3)\sin\omega t - \tfrac{1}{2}bV^2\cos 2\omega t + \tfrac{1}{4}cV^3\sin^3\omega t.$$

It is clear from these equations that even-order harmonics cancel out but odd-order harmonics do not and this means that the Class B push-pull amplifier introduces mainly third-harmonic distortion.

The bias arrangements for the output transistors given so far will not compensate for any changes in the base–emitter voltages that may occur because of temperature variations. To keep the biasing of the output transistors correct, most push-pull amplifiers employ a V_{BE} *multiplier* to derive the bias voltages. The circuit of a V_{BE} multiplier is given by Fig. 5.7 and it can directly replace resistor R_1 in the previous circuits. The transistor T_4 is biased to give the wanted bias voltages for the output transistors. The voltage drop from the base of T_2, the emitters of both T_2 and T_3, and the base of T_3 will be equal to the voltage drop across the V_{BE} multiplier, given by

$$V_{BE4}\left(1 + \frac{R_8}{R_9}\right)$$

and both will vary similarly with any change in temperature.

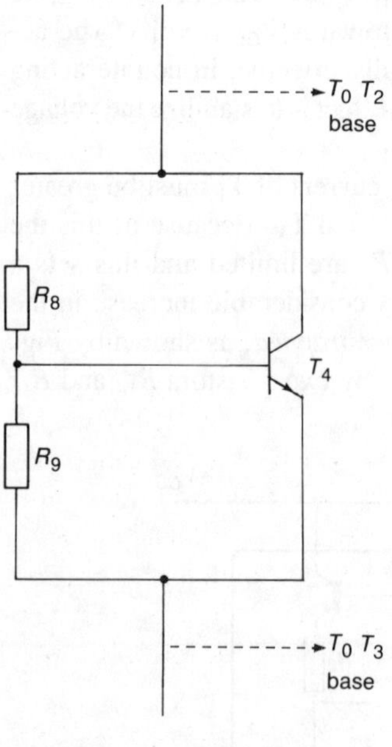

Fig. 5.7 V_{BE} multiplier.

Design of Class B Push-pull Amplifiers

The design of an amplifier of the kind shown in Fig. 5.5 starts from the specification of the wanted output power to a given value of load resistance. Suppose that the amplifier is to provide 10 W output power to an 8 Ω load over a bandwidth of 50 Hz to 20 kHz.

Then, $10 = V^2/8$ or $V = \sqrt{80} = 8.94 \simeq 9$ V and the peak output signal voltage is $\sqrt{2} \times 9$ or 12.7 V. The peak load current will then be $12.7/8 \simeq 1.6$ A. The quiescent potential at the junction of the emitter resistances R_3 and R_4 should be equal to $V_{CC}/2$ volts. Hence the maximum voltage at this point will be $12.7 + V_{CC}/2$. There will then be a near saturation voltage drop across the ON transistor, say 2 V, plus about 0.5 V drop across the emitter resistor. Therefore, $V_{CC} = 12.7 + 2.5 + V_{CC}/2$ or $V_{CC} = 30.4$ V, say 30 V.

The output transistors T_2 and T_3 must have ratings in excess of $V_{CE(max)} = 15$ V, $I_{c(max)} = 1.6$ A and $P_c = 0.4 \times 5 = 2$ W. A suitable device should be selected after consulting manufacturers' data sheets. The readily available and cheap BD131/2 could be chosen; these devices have maximum ratings at 25 °C ambient temperature of $V_{CEO} = 45$ V, $V_{CBO} = 70$ V, $h_{FE} = 20$(min), $P_T = 15$ W (at 60° C), and $f_t = 60$ MHz.

The peak base current to the output transistors should then be at most $1.6/20 = 0.080$ A. The two emitter resistors should each drop a d.c. voltage of about 0.5 V to ensure adequate thermal stability. The d.c. current of each transistor is $1.6/\pi = 0.51$ A and so $R_3 = R_4 = 0.5/0.51 \simeq 1$ Ω.

The d.c. voltage at the junction of R_3 and R_4 is $V_{CC}/2 = 15$ V and so the d.c. voltage at the base of T_2 should be equal to $15 + 0.7 + 0.5 = 16.2$ V. The voltage at the base of T_3 is equal to $15 - 0.7 - 0.5 = 13.8$ V.

The quiescent current of the driver transistor T_1 must be larger than the peak base current of either output transistor, say 100 mA. Then

$$R_1 = \frac{30 - 16.2}{0.1} = 138 \ \Omega, \text{ and } R_2 = \frac{16.2 - 13.8}{0.1} = 24 \ \Omega.$$

T_1 must be a transistor that is capable of passing a quiescent current of 100 mA and with a power rating of better than $0.1 \times 16.2 = 1.62$ W. A suitable device can be selected and its h_{FE} noted; potential problems may then arise with the voltage gain of this stage since its collector load resistance is so low, only $138 + 24 = 162$ Ω. It may well prove necessary to bootstrap R_1 in the manner shown in Fig. 5.6. Alternatively, the output transistors T_2 and T_3 can be replaced by Darlington pairs; the base drive requirement will then be much smaller and this will lead to considerably higher values for R_1 and R_2.

The output capacitor C_2 should be chosen to set the required lower 3 dB frequency of the circuit. Thus

$$C_2 = \frac{1}{2\pi \times 50 \times 8} = 398 \ \mu F.$$

The input capacitor C_1 can then be given any value that will make its contribution to the low-frequency response negligible. A typical value would be 22 μF.

Integrated-circuit Power Amplifiers

A large number of integrated-circuit power amplifiers are presently available from various manufacturers, giving output powers ranging from a few milliwatts to several watts. The most important parameters of such a device are its gain, input impedance, output impedance, quiescent supply current, supply voltage, bandwidth, distortion, sensitivity, and its maximum internal power dissipation. Information on these parameters and suggested circuits are provided by the manufacturer's data sheets. Some ICs must be mounted on a suitable heat sink and some are provided with short-circuit and/or thermal protection. The specific IC selected for a particular application depends upon the relative weights placed on the above parameters, the cost, and, of course, the ready availability from a convenient supplier. Some power amplifier ICs incorporate a pre-amplifier stage and/or an integral heat sink. Figure 5.8 shows the data sheet for the LM380 audio power amplifier (courtesy of National Semiconductor).

Both the input terminals of the IC are internally biased by a 150 kΩ resistance to earth and this allows any earth-referenced source to be directly connected to either input terminal. The unused input terminal can either be left disconnected or it can be connected to earth, either directly or via a resistor or a capacitor. When the inverting input is used it is best if the non-inverting pin is connected to earth to minimize noise pickup.

The basic circuit of a non-inverting circuit is given in Fig. 5.9(a). Should a variable gain be required it can be obtained by the use of an input potential divider as shown. For a high input resistance the alternative arrangement of Fig. 5.9(b) may be used, here the potential divider is connected between the two input terminals to give a differential input. If the power supply causes hum a capacitor C_2 can be connected from pin 1 to earth, of typical value 4.7 μF, and another capacitor C_3 connected across the supply (470 μF), these capacitors are both shown in Fig. 5.9(b).

The voltage gain of the circuit can also be reduced by the application of n.f.b. The output terminal 8 can be connected to the inverting input terminal 6 by a series resistor−capacitor circuit (R_2C_4). The voltage gain then obtained is equal to $A_{v(f)} = 1 + R_2/150$ kΩ. C_4 is necessary to prevent direct current flowing through the feedback path and its value affects the low-frequency response of the circuit, causing 3 dB fall in gain at a frequency f_1, where $f_1 = 1/(2\pi C_4 R_2)$.

Another example of an IC power amplifier is given by Fig. 5.10 which uses the TBA810. This IC, made by several manufacturers, can provide up to 7 W into a 4 Ω load with 10% distortion. Because of the high power output great care is needed in the layout to avoid instability problems. C_4, C_6, C_7 and R_4 provide high-frequency stability. C_3 bypasses supply voltage ripple to earth, C_8/C_9 are decoupling components, and C_1/C_{10} couple the amplifier to its source and its load respectively. R_3 and C_5 are bootstrap components while C_2R_2 are feedback components that establish the gain of the amplifier.

The characteristics of the TBA810 are:

National Semiconductor
LM380 Audio Power Amplifier
General Description

The LM380 is a power audio amplifier for consumer application. In order to hold system cost to a minimum, gain is internally fixed at 34 dB. A unique input stage allows inputs to be ground referenced. The output is automatically self centering to one half the supply voltage.

The output is short circuit proof with internal thermal limiting. The package outline is standard dual-in-line. A copper lead frame is used with the center three pins on either side comprising a heat sink. This makes the device easy to use in standard p-c layout.

Uses include simple phonograph amplifiers, intercoms, line drivers, teaching machine outputs, alarms, ultrasonic drivers, TV sound systems, AM-FM radio, small servo drivers, power converters, etc.

Features

- Wide supply voltage range
- Low quiescent power drain
- Voltage gain fixed at 50
- High peak current capability
- Input referenced to GND
- High input impedance
- Low distortion
- Quiescent output voltage is at one-half of the supply voltage
- Standard dual-in-line package

Connection Diagrams (Dual-in-line packages, top view)

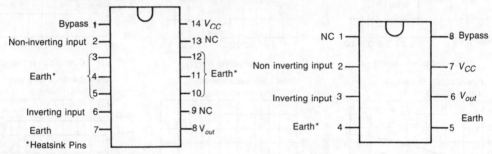

Absolute Maximum Ratings

Supply voltage	22 V
Peak current	1.3 A
Package Dissipation 14-pin DIP (Notes 6 and 7)Operating temperature range	10 W
Input voltage	± 0.5 V
Storage temperature	$-65\,^{\circ}\text{C}$ to $+150\,^{\circ}\text{C}$
Operating temperature	$0\,^{\circ}\text{C}$ to $+70\,^{\circ}\text{C}$
Junction temperature	$+150\,^{\circ}\text{C}$
Lead temperature (soldering, 10 s)	$+300\,^{\circ}\text{C}$

Electrical Characteristics (Note 1)

Parameter	Symbol	Conditions	Min	Typ	Max	Units
Output power	$P_{out(rms)}$	(Notes 3,4)$R_L = 8\,\Omega$, t.h.d. $= 3\%$	2.5			W
Gain	A_V		40	50	60	V/V
Output voltage swing	V_{out}	$R_L = 8\,\Omega$		14		$V_{p\text{-}p}$
Input resistance	Z_{in}			150k		Ω
Total harmonic distortion	t.h.d.	(Note 4,5)		0.2		%
Power supply rejection ratio	p.s.r.r.	(Note 2)		38		dB
Supply voltage	V_{CC}	(Note 8)	10		22	V
Bandwidth	BW	$P_{out} = 2$ W, $R_L = 8\,\Omega$		100k		Hz
Quiescent supply current	I_Q			7	25	mA
Quiescent output voltage	V_{outQ}		8	9.0	10	V
Bias current	I_{bias}	Inputs floating		100		nA
Short circuit current	I_{sc}			1.3		A

Fig. 5.8   Data sheet for LM 380 audio-frequency power amplifier (*courtesy National Semiconductor (UK) Ltd*).

Note 1: V_{CC} = 18 V and T_A = 25 °C unless otherwise specified.
Note 2: Rejection ratio referred to the output with C_{bypass} = 5 μF.
Note 3: With device Pins 3,4,5,10,11,12 soldered into a 1/16" epoxy glass board with 2 ounce copper foil with a minimum surface of 6 square inches.
Note 4: If oscillation exists under some load conditions, add 2.7 Ω and 0.1 μF series network from Pin 8 to earth.
Note 5: C_{bypass} = 0.47 μF on Pin 1.
Note 6: The maximum junction temperature of the LM380 is 150 °C.
Note 7: The package is to be derated at 12° C/W junction to heat sink pins.
Note 8: Can select for 8 V operation.

Typical Performance Characteristics

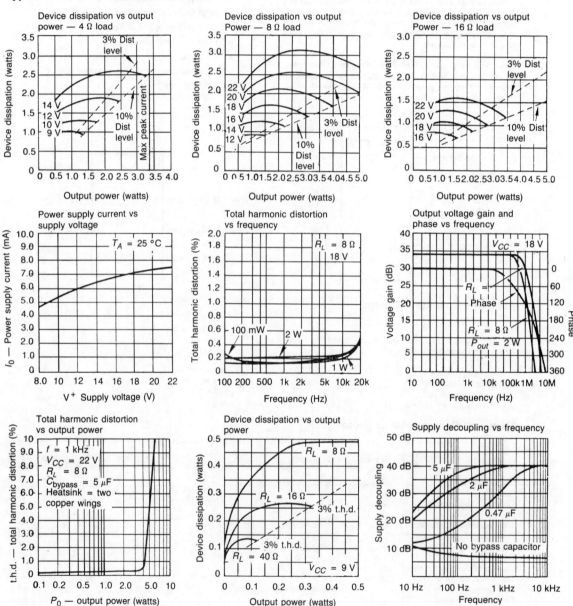

Fig. 5.8 cont.

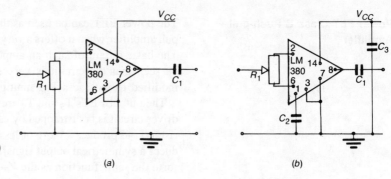

Typical values

$R_1 = 10\ \text{k}\Omega \quad C_1 = 470\ \mu\text{F}$
$C_2 = 4.7\ \mu\text{F}$

Fig. 5.9 LM 380 non-inverting power amplifier (a) basic circuit, (b) high-impedance circuit.

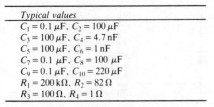

Typical values
$C_1 = 0.1\ \mu\text{F}, C_2 = 100\ \mu\text{F}$
$C_3 = 100\ \mu\text{F}, C_4 = 4.7\ \text{nF}$
$C_5 = 100\ \mu\text{F}, C_6 = 1\ \text{nF}$
$C_7 = 0.1\ \mu\text{F}, C_8 = 100\ \mu\text{F}$
$C_9 = 0.1\ \mu\text{F}, C_{10} = 220\ \mu\text{F}$
$R_1 = 200\ \text{k}\Omega, R_2 = 82\ \Omega$
$R_3 = 100\ \Omega, R_4 = 1\ \Omega$

Fig. 5.10 TBA 810 power amplifier 1 supply voltage: 2 2/3 NC: 4 bootstrap: 5 compensation: 6 feedback: 7 bypass: 8 input: 9 input earth: 10 output earth: 11 NC: 12 output.

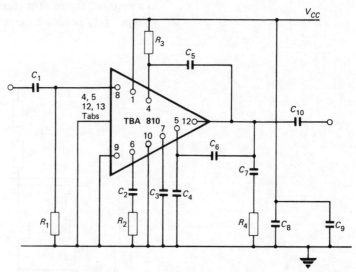

Supply voltage V_{CC} = 4 V – 20 V
Output power P_{out} 2.5 W – 7 W (depending upon V_{CC})
t.h.d. for P_{out} up to 3 W = 0.3%
Closed-loop voltage gain = 37 dB
Input resistance = 5 MΩ
Efficiency = 70%.

Table 5.1 lists some of the more important parameters of a number of other IC power amplifiers.

Table 5.1 IC power amplifiers

I.C.	Power output (W)	Load resistance (Ω)	Voltage gain (dB)	Input resistance
TDA 2610	4.5	15	37	45 kΩ
LM 389	0.325	8	26	50 kΩ
TBA 820	1.6	8	34	5 MΩ
CA 3131	5	8	48	200 kΩ
TDA 2030	17	4	40	5 MΩ
TDA 800	5	16	42	5 MΩ

Power FET Class B Push-pull Amplifier

The power FET can be used as the output device in a Class B push-pull amplifier when it offers a very good high-frequency performance. The basic circuit of such an amplifier is shown in Fig. 5.11; it can be seen that the circuit is very similar to that given in Fig. 5.7, modified to include a V_{BE} multiplier in place of R_2.

The output FETs T_3 and T_4 are operated as source followers. The driver circuit is bootstrapped by capacitor C_1; this allows driver transistor T_1 to produce a more or less constant d.c. current and also produces a symmetrical output signal. T_2 and its associated resistors perform the same function as the V_{BE} multiplier of Fig. 5.7, R_3 is often a variable component to allow the output quiescent current to be set to a required figure. The diode D_1 is needed to ensure that the gate voltage of T_4 never becomes more positive than the supply voltage V_{DD}.

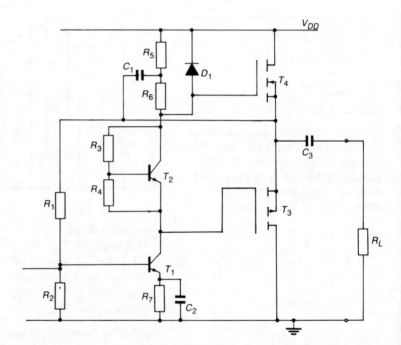

Fig. 5.11 Power FET amplifier.

6 Sinusoidal Oscillators

An oscillator is an electronic circuit whose function is to produce an alternating e.m.f. of a particular frequency and waveform. This chapter will be concerned solely with oscillators which generate an output of sinusoidal waveform, and other waveform generators will be discussed in Chapter 7.

All types of sinusoidal oscillator rely upon the application of positive feedback to a circuit that is capable of providing amplification, and they differ from one another mainly in the ways in which the feedback is applied. The basic block diagram of an oscillator is given in Fig. 6.1. When the oscillator is first switched on, a surge of current flows in the frequency-determining network and produces a voltage at the required frequency of oscillation across it. A fraction of this voltage is fed back, via the feedback network, to the input terminals of the amplifier and is then amplified to reappear across the frequency-determining network. A fraction of this larger voltage is fed back to the input, in phase with the input voltage, and is further amplified and so on. In this way the amplitude of the signal voltage builds up until the onset of non-linearity in the operation of the amplifier which reduces the loop gain to unity.

The frequency-determining network may consist of an *LC* circuit, or an *RC* circuit, or a piezo-electric crystal. The amplifying device may be a bipolar transistor, or a FET, or an operational amplifier.

The important characteristics of an oscillator are its (*a*) frequency, or range of frequencies (if available), of operation, (*b*) frequency stability, (*c*) amplitude stability, and (*d*) percentage distortion of its output waveform.

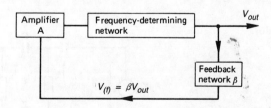

Fig. 6.1 The principle of an oscillator.

The Generalized Oscillator

Referring to Fig. 6.1 the input voltage to the amplifying section is

$$V_{in} = \beta V_{out} = \beta A_v V_{in}$$

and so $V_{in}(1 - \beta A_v) = 0.$

The input voltage cannot be zero as an output voltage does exist and therefore $(1 - \beta A_v)$ must be zero. Hence $\beta A_v = 1$.

In general, both the gain A_v *and* the feedback factor β are complex and hence

$$|\beta| \angle \phi \cdot |A_v| \angle \theta = 1 \angle 0°$$

$$|\beta A_v| \angle \phi + \theta = 1 \angle 0°. \tag{6.1}$$

Equation (6.1) states that the two necessary requirements for oscillation to take place are that

the *loop gain* $|\beta A_v|$ *must be unity*
and *the loop phase shift* $\phi + \theta$ *must be 360°*.

This means that the frequency of oscillation of an oscillator circuit can be determined by equating the j parts of equation (6.1) to zero.

Equation (6.1) is often known as the *Barkhausen criterion*. However, it is not practical to make an oscillator with a loop gain of exactly unity, and practical circuits have a loop gain which is slightly greater than unity and rely on non-linearity to limit the amplitude of the oscillations.

Resistance–capacitance Oscillators

Audio-frequency oscillators most often use a resistance–capacitance network to obtain the loop phase shift of 360° necessary for oscillations to take place. One type of *RC* oscillator uses an op-amp and an *RC* ladder network, and its circuit is given in Fig. 6.2.

Because of the virtual earth the resistor R_1 connected to the inverting input of the op-amp is effectively connected to earth. The resistors R_1, R_3 and R_4 are of equal value, say R, and so are the three capacitors, say $C = C_1 = C_2 = C_3$. The equivalent circuit of the oscillator is shown by Fig. 6.3. From Fig. 6.3,

$$A_v V_{in} = I_1(R - j/\omega C) - I_2 R \tag{6.2}$$

$$0 = I_2(2R - j/\omega C) - I_1 R - I_3 R \tag{6.3}$$

$$0 = I_3(2R - j/\omega C) - I_2 R. \tag{6.4}$$

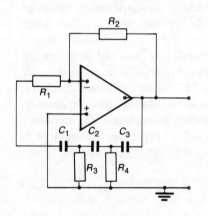

Fig. 6.2 *RC* oscillator.

Fig. 6.3 Equivalent circuit of Fig. 6.2.

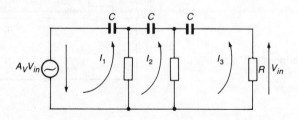

From equation (6.4),

$$I_2 = \frac{I_3(2R - j/\omega C)}{R} = I_3(2 - j/\omega CR). \tag{6.5}$$

From equation (6.3),

$$I_1 = I_2(2 - j/\omega CR) - I_3 = I_3(2 - j/\omega CR)^2 - I_3. \tag{6.6}$$

Substituting (6.5) and (6.6) into equation (6.2) gives

$$A_v V_{in} = [I_3(2 - j/\omega CR)^2 - I_3](R - j/\omega C) - RI_3(2 - j/\omega CR)$$

$$= I_3\left[\left(4 - \frac{j4}{\omega CR} - \frac{1}{\omega^2 C^2 R^2} - 1\right)(R - j/\omega C) - 2R + \frac{j}{\omega C}\right]$$

$$= I_3\left[R - \frac{5}{\omega^2 C^2 R} + j\left(\frac{1}{\omega^3 C^3 R^2} - \frac{6}{\omega C}\right)\right].$$

Now, $V_{in} = I_3 R$ and $A_v = A_v V_{in}/V_{in}$, so that

$$A_v = 1 - \frac{5}{\omega^2 C^2 R^2} + j\left(\frac{1}{\omega^3 C^3 R^3} - \frac{6}{\omega CR}\right). \tag{6.7}$$

The circuit will oscillate at the frequency f_0 at which equation (6.7) is wholly real, i.e. at which the sum of the j terms is zero. Therefore

$$\frac{1}{\omega_0^3 C^3 R^3} = \frac{6}{\omega_0 CR} \quad \text{or} \quad \omega_0^2 = \frac{1}{6C^2 R^2}$$

$$\omega_0 = \frac{1}{\sqrt{6}CR} \quad \text{and} \quad f_0 = \frac{1}{2\pi\sqrt{6}CR}. \tag{6.8}$$

At this frequency,

$$A_v = 1 - \frac{5 \times 6C^2 R^2}{C^2 R^2} = 1 - 30, \quad \text{or} \quad A_v = -29.$$

This means that the voltage gain of the circuit, $A_v = -R_2/R_1$, must be slightly in excess of 29, say 31.

The circuit is capable of operation over a range of frequencies from a few Hertz to more than 100 kHz, the maximum frequency being limited by the slew rate of the op-amp. The frequency of oscillation can be varied by changing any one, or more, of the frequency-determining elements. Very often the three capacitors are ganged and varied simultaneously.

The design of the circuit can proceed as follows. The maximum output voltage V_{out} is about $V_{CC} - 1$ volts so that the input voltage is $(V_{CC} - 1)/31$. The input current I_{in} should be several times n larger than the maximum input bias current $I_{B(max)}$. Hence, $R_1 = (V_{CC} - 1)/31nI_{B(max)}$ and $R_2 = 31R_1$. If $R_3 = R_4 = R_1$ the value of $C_1 = C_2 = C_3$ can be calculated from equation (6.8).

Example 6.1

Design an *RC* oscillator using a 741 op-amp to have a frequency of 2 kHz. Use a power supply voltage of ± 12 V.

Solution

The maximum input voltage is $V_{in} = 11/31 = 0.355$ V. The 741 op-amp has $I_{B(max)} = 500$ nA, so try $n = 100$ to give $I_{in} = 50$ μA. Then $R_1 = 0.355/(50 \times 10^{-6}) = 7.1$ kΩ. The nearest preferred higher value is 7.5 kΩ. Then $R_2 = 31 \times 7.1 = 220$ kΩ. (Now $A_v = 220/7.5 = 29.3$ which is only marginally all right.) Try instead 6.8 kΩ as the preferred value for the 7.1 kΩ resistance. Then $A_v = 220/6.8 = 32.4$. Lastly, $C = 1/(2\pi \times 2000 \times \sqrt{6 \times 10^3}) = 4.8$ nF. The nearest preferred value is 4.7 nF; using this gives f_0 as 2033 Hz.

Wien Bridge Oscillator

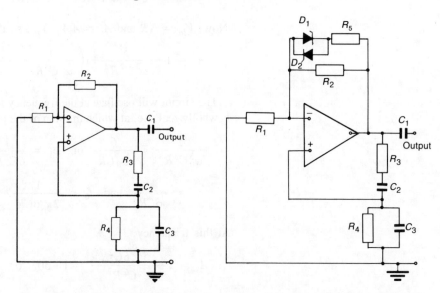

Fig. 6.4 Wien bridge oscillator.

Fig. 6.5 Use of Zener diodes to limit oscillation amplitude.

Another kind of *RC* oscillator is known as the Wien bridge circuit and is shown in Fig. 6.4 The voltage applied to the non-inverting terminal is

$$V_+ = \cfrac{\cfrac{V_{out}R_4}{1+j\omega C_3 R_4}}{\cfrac{R_4}{1+j\omega C_3 R_4}+R_3+\cfrac{1}{j\omega C_2}}$$

$$= \cfrac{V_{out}R_4}{\cfrac{C_3 R_4}{C_2}+R_3+R_4+j\left(\omega R_3 R_4 C_3 - \cfrac{1}{\omega C_2}\right)}.$$

V_+ will be in phase with V_{out}, and the circuit will oscillate, when the j terms sum to zero, i.e. when

$$\omega_0 R_3 R_4 C_3 = 1/\omega_0 C_2$$

$$\omega_0 = \frac{1}{\sqrt{[R_3 R_4 C_2 C_3]}} \quad \text{and} \quad f_0 = \frac{1}{2\pi\sqrt{[R_3 R_4 C_2 C_3]}} \text{Hz.} \quad (6.9)$$

At this frequency

$$V_{out}/V_+ = 1 + \frac{R_3}{R_4} + \frac{C_3}{C_2} \quad (6.10)$$

and this is the amplifier gain necessary for a loop gain of unity. If $R_3 = R_4$ and $C_2 = C_3$ the amplifier gain needed is 3 and this is easily obtained by making $R_2 = 2R_1$.

The oscillations build up in amplitude until amplifier non-linearities cause the loop gain to fall to unity. Frequency variation can be achieved by the simultaneous variation of the ganged capacitors C_2 and C_3. If required, range switching can be obtained by switching in new values for R_3 and R_4 (usually equal).

The amplitude of the oscillations is determined by the amount by which the loop gain of the circuit is in excess of unity. To prevent the op-amp driving well into saturation on both half cycles the loop gain must be limited in some way. The negative-feedback path, which sets the gain of the circuit, can contain an element whose resistance is temperature, and hence power-dissipation, sensitive. Two possibilities exist: (a) an element with a negative temperature coefficient, such as a thermistor, can replace R_2 in Fig. 6.4; and (b) a positive temperature-coefficient component, such as a lamp, can replace R_1. For case (a) the oscillation amplitude builds up, the voltage across R_2 increases also and the higher power dissipation raises the temperature of, and therefore reduces the resistance of R_2. The reduction in the ratio R_2/R_1 reduces the gain of the amplifier and so limits the oscillation amplitude.

If the thermistor has a maximum permitted power dissipation of P watts and oscillation amplitude is to be V_{out} volts, then, since $R_2 = 2R_1$, the voltage across the thermistor should be $2V_{out}/3$ volts. The current I flowing in both R_1 and the thermistor is then $I = P/(2V_{out}/3)$ amps, and hence $R_1 = (V_{out}/3)/I$ Ω. The value of R_1 should then be chosen to be slightly larger than the calculated value in order to keep the thermistor's dissipation below its maximum value. The thermal time constant of the temperature-sensitive component must be long compared with the periodic time of the oscillation waveform so that for a constant amplitude the device will act like a resistor.

Both of these methods suffer from the disadvantages of extra power dissipation and a slow response time, and an alternative is to include a non-linear network in the n.f.b. loop. One method is shown in Fig. 6.5. Initially, the two Zener diodes are non-conducting and the gain of the circuit is $1 + R_2/R_1$. When the oscillation amplitude increases the point will be reached where the diodes conduct and the gain is reduced.

The Wien oscillator is preferred to the ladder RC oscillator when a variable-frequency output is wanted since there are fewer components to be simultaneously controlled.

Design of an Op-amp Wien Oscillator

The maximum output voltage will be equal to $(V_{CC}-1)$ volts and the current flowing in R_2 should be many n times larger than the maximum input bias current $I_{B(max)}$. Hence $R_1 + R_2 = (V_{CC}-1)/nI_{B(max)}$. If $R_1 = R_2$ and $C_2 = C_3$ then $R_2 = 2R_1$ and $R_1 = (V_{CC}-1)/3nI_{B(max)}$. The value of n is not critical and can be selected to give convenient values for R_1 and R_2: Lastly, a convenient value can be chosen for either R_3 and R_4, or for C_2 and C_3 and the other value can be calculated using equation (6.9). The selected op-amp should, of course, have sufficiently high values of full-power bandwidth and slew rate to be able to provide the desired output frequency and voltage.

Example 6.2

Design an op-amp Wien bridge oscillator to operate at 4 kHz. Use a 741 op-amp and a supply voltage of ± 12 V.

Solution
The 741 op-amp has a maximum input bias current of 500 nA. If n is chosen as 100, $I_2 = 50\ \mu A$ and $R_1 + R_2 = 11/(50 \times 10^{-6}) = 220$ kΩ. Then $R_1 = 73.3$ kΩ and $R_2 = 146.6$ kΩ. Using the nearest preferred values $R_1 = 75$ kΩ and $R_2 = 150$ kΩ. These rather large values may lead to some noise problems, and another value for n could be tried. Suppose $n = 1000$ then $I_2 = 500\ \mu A$ and $R_1 + R_2 = 11/(500 \times 10^{-6}) = 22$ kΩ, and $R_1 = 7.5$ kΩ and $R_2 = 15$ kΩ. Other values of n will obviously lead to other pairs of values for R_1 and R_2.

If $R_3 = R_4 = 7.5$ kΩ then $C_2 = C_3 = 1/(2\pi \times 4000 \times 7.5 \times 10^3)$ $= 5.3$ nF. If the nearest preferred value is used, i.e. 4.7 nF, the frequency of oscillation will be 4515 Hz. If, on the other hand, C_2 is selected to be 4.7 nF initially, R_3 will then be calculated to be 8465 Ω, and using the nearest preferred value of 8.2 kΩ will give a frequency of 4130 Hz. The control of the oscillation amplitude must now be considered.

(a) If a thermistor is to be used in place of R_2 it will have a voltage drop of $22/3 = 7.33$ V across it and will pass a current of $(11 - 7.33)/7.5 \times 10^3 \simeq 0.5$ mA. The thermistor must therefore have a rated power dissipation in excess of $7.33 \times 0.5 \simeq 3.6$ mW. A suitable device can then be selected from manufacturer's data.

(b) If the diode control circuit of Fig. 6.5 is to be used then the gain is to vary from say 2.7 to 3.3. Hence

$$3.3 = 1 + R_2/7.5 \quad \text{or} \quad R_2 = 17.25 \text{ k}\Omega$$

and

$$2.7 = 1 + R/7.5 \quad \text{or} \quad R = 12.75 \text{ k}\Omega.$$

$$R = R_2 R_5/(R_2 + R_5)$$

so that

$$R_5 = 48.88 \text{ k}\Omega$$

If the nearest preferred values are used, $R_2 = 18 \text{ k}\Omega$ and $R_5 = 47 \text{ k}\Omega$ or $51 \text{ k}\Omega$. These values give $A_v = 1 + 18/7.5 = 3.4$; and either $A_v = 1 + 13/7.5 = 2.73$, or $A_v = 1 + 13.3/7.5 = 2.77$.

Inductance–Capacitance Oscillators

Resistance–capacitance oscillators are not suited to use at frequencies in excess of some hundreds of kilohertz because the required values of R and C become impracticably small. For the generation of higher frequencies, the frequency-determining network is provided by an inductance–capacitance network.

A number of different configurations are possible but perhaps the most popular are the tuned-collector oscillator, the Hartley oscillator, and Colpitts oscillator.

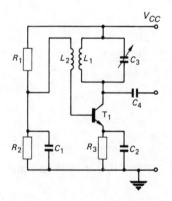

Fig. 6.6 Tuned-collector oscillator.

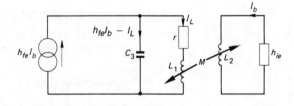

Fig. 6.7 *h*-parameter equivalent circuit of Fig. 6.6.

Tuned-collector Oscillator

Figure 6.6. shows the circuit of a tuned-collector oscillator in which R_1, R_2 and R_3 are bias components while C_1 and C_2 are decoupling components. Variable capacitor C_3 tunes the circuit to the desired frequency of oscillation and C_4 is a d.c. blocker. The action of the circuit is as follows. When the collector supply voltage is first switched on, the resulting surge in the d.c. current causes a minute oscillatory current to flow in the tuned circuit. This current flows in the inductor L_1 and induces a voltage at the same frequency into the inductor L_2. This voltage is then applied to the base of the transistor. The transistor introduces a 180° phase shift between its base and its collector terminals, and the mutual inductance between L_1 and L_2 must be such that the loop phase shift is zero. The amplified voltage causes a larger oscillatory current to flow in L_1 and a larger e.m.f. is induced into L_2 and so on. Provided the loop gain is greater than unity, the amplitude of the oscillations builds up until the point is reached where the transistor is driven into cut-off and saturation. The loop gain is then reduced to unity and the oscillation amplitude remains constant.

The derivation of the maintenance conditions required for steady-state sinusoidal oscillations to occur in an LC oscillator will be based upon the simplified h-parameter a.c. equivalent circuit. If the oscillation frequency is well above the audio range, the h-parameter circuit is not really valid and the hybrid π circuit should be used. The results obtained in the following pages are approximately correct and are commonly used as a basis for design.

The equivalent circuit of the tuned collector oscillator is shown in Fig. 6.7. Applying Kirchhoff's law to the collector circuit

$$0 = I_L(r+j\omega L_1) - j\omega M I_b - (h_{fe}I_b - I_L)\frac{1}{j\omega C_3}$$

$$I_L(j/\omega C_3 - r - j\omega L_1) = I_b(-j\omega M + jh_{fe}/\omega C_3). \tag{6.11}$$

And, from the base circuit,

$$0 = -I_b(h_{ie}+j\omega L_2) + j\omega M I_L$$

$$j\omega M I_L = I_b(h_{ie}+j\omega L_2). \tag{6.12}$$

Divide equation (6.11) by (6.12),

$$\frac{-r+j(1/\omega C_3 - \omega L_1)}{j\omega M} = \frac{j(-\omega M + h_{fe}/\omega C_3)}{h_{ie}+j\omega L_2}$$

and cross-multiplying

$$-rh_{ie} - j\omega r L_2 + jh_{ie}(1/\omega C_3 - \omega L_1) - L_2/C_3 + \omega^2 L_1 L_2$$
$$= \omega^2 M^2 - h_{fe}M/C_3. \tag{6.13}$$

For oscillations to occur the j part of equation (6.13) must be equal to zero, hence

$$-\omega_0 r L_2 + h_{ie}(1/\omega_0 C_3 - \omega_0 L_1) = 0$$

$$-\omega_0^2 r L_2 C_3 + h_{ie} - h_{ie}\omega_0^2 L_1 C_3 = 0$$

$$\omega_0^2 = h_{ie}/(L_1 C_3 h_{ie} + L_2 C_3 r).$$

Therefore,

$$f_0 = \frac{1}{2\pi\sqrt{[L_1 C_3 + (L_2 C_3 r/h_{ie})]}} \tag{6.14}$$

$$f_0 \simeq \frac{1}{2\pi\sqrt{(L_1 C_3)}} \quad \text{if} \quad h_{ie} \gg r. \tag{6.15}$$

Equating the real parts of equation (6.13) gives

$$-rh_{ie}-\frac{L_2}{C_3}+\omega_0^2 L_1 L_2 = \omega_0^2 M^2 - h_{fe}\frac{M}{C_3}$$

$$h_{fe} = \frac{1}{M}[rh_{ie}C_3+L_2+\omega_0^2 C_3(M^2-L_1L_2)]$$

$$\simeq \frac{rh_{ie}C_3}{M}+\frac{L_2}{M}+\frac{C_3}{L_1 C_3 M}(M^2-L_1L_2)$$

$$h_{fe(min)} = \frac{rh_{ie}C_3}{M}+\frac{M}{L_1}. \tag{6.16}$$

Equations (6.14) and (6.16) ignore the effects of h_{oe} and the impedance into which the oscillator output is delivered. These have the effect of decreasing the calculated value for ω_0 and increasing the calculated required minimum value of h_{fe}.

Colpitts and Hartley Oscillators

Two other types of LC oscillator that are often used in electronic equipment are the Colpitts and Hartley oscillators. These oscillators can be represented by the generalized diagram shown in Fig. 6.8 in which X_1 is the reactance connected between input and output, X_2 is the reactance between the input and earth, and X_3 is the reactance between the output and earth.

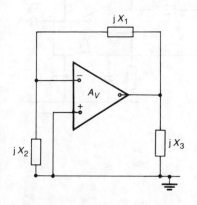

Fig. 6.8 Three-impedance oscillator.

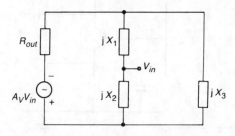

Fig. 6.9 Equivalent circuit of Fig. 6.8.

The equivalent circuit of Fig. 6.8 is given by Fig. 6.9. The load into which the op-amp works is equal to

$$Z_L = \frac{jX_3(jX_1+jX_2)}{jX_1+jX_2+jX_3}$$

and so the voltage gain without feedback is

$$\frac{-(A_v Z_L)}{(R_{out}+Z_L)} = \frac{A_v X_3(X_1+X_2)}{jR_{out}(X_1+X_2+X_3)-X_3(X_1+X_2)}.$$

The feedback factor β is equal to $X_2/(X_1+X_2)$ and so the loop gain is $A_vX_2X_3/[jR_{out}(X_1+X_2+X_3)-X_3(X_1+X_2)]$. For oscillations to occur the loop phase shift must be $360°$ or $0°$ and hence the j terms must sum to zero. Thus

$$X_1+X_2+X_3 = 0. \tag{6.17}$$

This means that one of the three reactances must be of the opposite sign to the other two reactances. The loop gain is then unity and is

$$1 = \frac{-A_vX_2}{X_1+X_2} = \frac{A_vX_2}{X_3}.$$

Since this must be positive, both X_2 and X_3 must have the same sign and hence X_1 must be of the opposite sign.

For a *Colpitts oscillator*, X_2 and X_3 are capacitive reactances and X_1 is an inductive reactance.

For a *Hartley oscillator*, X_2 and X_3 are inductive and X_1 is capacitive.

Colpitts Oscillator

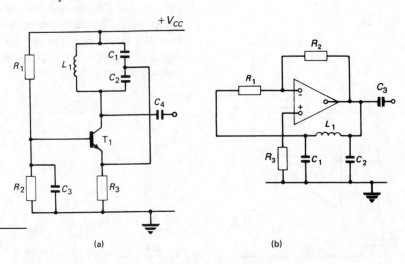

Typical values
(a) $R_1 = 56\,k\Omega$, $R_2 = 12\,k\Omega$
$R_3 = 1\,k\Omega$, $C_1 = 0.47\,\mu F$
$C_2 = 22\,nF$, $C_3 = 47\,nF$
$C_4 = 0.1\,\mu F$, $L_1 = 1\,mH$
(b) $R_1 = 10\,k\Omega$, $R_2 = 56\,k\Omega$
$R_3 = 8.2\,k\Omega$, $C_1 = 22\,nF$
$C_2 = 10\,nF$, $C_3 = 0.1\,\mu F$
$L_1 = 20\,mH$

Fig. 6.10 Colpitts oscillators.

(a)   (b)

The circuit diagrams of op-amp and transistor Colpitts oscillators are shown in Fig. 6.10. The frequency of oscillation is easily obtained from equation (6.17), by putting $jX_1 = j\omega L_1$, $jX_2 = 1/j\omega C_1$, $jX_3 = 1/j\omega C_2$, to give

$$\frac{1}{j\omega_0 C_1} + \frac{1}{j\omega_0 C_2} + j\omega_0 L_1 = 0$$

$$\frac{1}{C_1} + \frac{1}{C_2} = \omega_0^2 L_1$$

$$\omega_0^2 = \frac{1}{L_1}\left(\frac{1}{C_1}+\frac{1}{C_2}\right) = \frac{1}{L_1}\left(\frac{C_1+C_2}{C_1 C_2}\right)$$

$$f_0 = \frac{1}{2\pi\sqrt{\left(L_1\dfrac{C_1 C_2}{C_1+C_2}\right)}}\text{Hz.} \qquad (6.18)$$

The minimum voltage gain A_v the op-amp must give is obtained from $1 = \dfrac{A_v X_2}{X_3}$, therefore

$$A_{v(min)} = X_3/X_2 = C_1/C_2. \qquad (6.19)$$

The analysis is less accurate for a transistor circuit because the relatively low input resistance h_{ie} of the device will shunt the reactance X_2.

Hartley Oscillator

Figure 6.11 shows the circuits of op-amp and transistor Hartley oscillators. Now, $X_1 = 1/j\omega C_1$, $X_2 = j\omega L_1$, and $X_3 = j\omega L_2$. Substituting into equation (6.17) gives

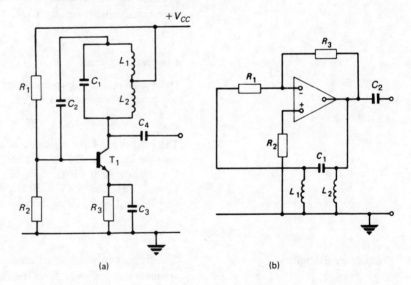

Typical values
(a) $R_1 = 75\,\text{k}\Omega$, $R_2 = 20\,\text{k}\Omega$
$R_3 = 1.5\,\text{k}\Omega$, $C_1 = 0.1\,\mu\text{F}$
$C_2 = 40 - 250\,\text{pF}$
$C_3 = 1\,\mu\text{F}$, $C_4 = 0.1\,\mu\text{F}$
$L_1 = L_2 = 125\,\mu\text{H}$
(b) $R_1 = 39\,\text{k}\Omega$, $R_2 = 150\,\text{k}\Omega$
$R_3 = 30\,\text{k}\Omega$, $C_1 = 0.02\,\mu\text{F}$
$C_2 = 0.1\,\mu\text{F}$, $L_1 = 8\,\text{mH}$
$L_2 = 5\,\text{mH}$

Fig. 6.11 Hartley oscillators.

(a)　　　　(b)

$$\frac{1}{j\omega_0 C}+j\omega_0 L_1+j\omega_0 L_2 = 0$$

$$\frac{1}{C_1} = \omega_0^2(L_1+L_2) \qquad \omega_0^2 = \frac{1}{C_1(L_1+L_2)}$$

$$f_0 = \frac{1}{2\pi\sqrt{[C_1(L_1+L_2)]}}. \qquad (6.20)$$

The loop gain is $A_v X_3/X_2$ and for this to be unity or greater, $A_{v(min)} = X_2/X_3 = L_1/L_2$.

The Hartley oscillator has two disadvantages compared with the Colpitts oscillator. Some mutual inductance may exist between the two inductors L_1 and L_1 and this will affect the frequency of oscillation in a way that is difficult to predict. Also, it is not easy to vary inductance values over a wide range. Usually, the Colpitts oscillator is employed, generally with variable capacitors although some circuits may use a variable inductance.

Design of Colpitts and Hartley Oscillators

The basic design procedure for the two oscillators are essentially the same and so only the Colpitts circuit is considered. The reactance of C_2 appears in parallel with the output resistance of the op-amp and so it should be several times, say ten, larger than it. Similarly, one side of the input resistor R_1 is at virtual earth and hence the value of R_1 should be many times greater than the reactance of C_1. The value of L_1 can be calculated from the expression for the oscillation frequency (6.18), and the value of R_2 can be found from the gain expression for an inverting amplifier. This should be larger than the minimum value given by C_1/C_2.

Example 6.3

Design an op-amp Colpitts oscillator to operate at 3 kHz. Use a 741 and a power supply voltage of ± 12 V.

Solution

The 741 has an output resistance of about 70 Ω so that the reactance of C_2 must be at least 700 Ω. If equal tuning capacitors are chosen then $C = 1/(2\pi \times 3000 \times 700) = 76$ nF, so that $C_1 = C_2 = 152$ nF. Then $L_1 = 37$ mH. Choose $R_1 = 20$ kΩ, which is several times larger than X_{C1}, then R_2 must be 22 kΩ or more.

Frequency Stability

The frequency stability of an oscillator is the amount by which its frequency drifts away from the desired value. It is desirable that the frequency drift should be small, and the maximum allowable change in frequency is usually specified as so many parts per million, e.g. ± 1 part in 10^6 would mean a maximum frequency drift of ± 1 Hz if the frequency of oscillation were 1 MHz but of $\pm$ 100 Hz if the frequency were 100 MHz

Factors that affect the frequency stability of an oscillator are (a) the load into which the oscillator works, (b) the parameters of the active device, (c) the supply voltage, and (d) the stability of the components used in the frequency-determining network.

Load on the Oscillator

The frequency of oscillation is not independent of the load into which the oscillatory power is delivered. If the load resistance should vary, the frequency of oscillation will not be stable. Variations in the external load can be effectively removed by connecting a buffer amplifier between the oscillator and its load.

Variations in the Supply Voltage

The parameters of an active device, such as the current or voltage gain and the input and output capacitances, are functions of the quiescent collector current and hence of the supply voltage. Any changes in the supply voltage will cause one or more parameters to vary and the oscillation frequency to drift. This cause of frequency instability can be minimized by the use of adequate power supply stabilization.

Circuit Components

Changes in the temperature of the circuit components will produce changes in inductance and capacitance and hence in the oscillation frequency. Steps that can be taken to minimize this effect include: keeping the frequency-determining components well clear of any heat-producing sources; and mounting the components inside a thermostatically controlled oven.

Crystal Oscillators

The best frequency stabilily that can be achieved with an LC oscillator is limited to about ± 100 parts in 10^6 per °C and if a better frequency stabililty is needed a crystal oscillator must be employed. A crystal oscillator is one in which the frequency-determining network is provided by a piezo-electric crystal.

Piezo-electric Crystals

A piezo-electric crystal is a material, such as quartz, having the property that, if subjected to a mechanical stress, a potential difference is developed across it, and if the stress is reversed, a p.d. of the opposite polarity is produced. Conversely, the application of a potential difference to a piezo-electric crystal causes the crystal to be stressed in a direction depending on the polarity of the applied voltage.

If a small, thin plate is cut from a crystal and two electrodes are vacuum deposited on to its faces, the plate will possess a particular natural frequency and, if an alternating voltage at this frequency is applied across the electrodes, the plate will vibrate vigorously. The natural frequency of a crystal depends upon its dimensions, the mode of vibration, and its original position or *cut* in the crystal. The important characteristics of a particular cut are its natural frequency and its temperature coefficient. One cut, the GT cut, has a negligible temperature coefficient over a temperature range of 0 °C to 100 °C. The AT cut has a temperature coefficient that varies from about +10 p.p.m./°C at 0 °C to 0 p.p.m. at 40 °C and about +20 p.p.m./°C at 90 °C. Crystal plates are available with fundamental natural frequencies from about 4 kHz up to about 30 MHz or so. For higher frequencies the required plate thickness is very small and the plate becomes very fragile; however, a crystal can be operated at an odd *overtone* frequency and such overtone operation raises the possible upper frequency to about 100 MHz. An overtone frequency is *not* exactly related to the fundamental frequency.

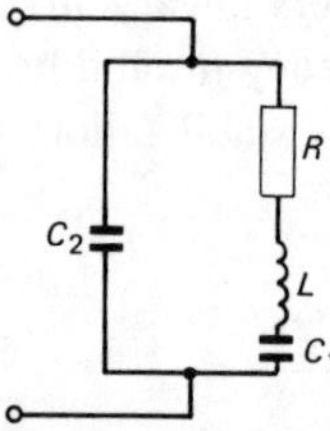

The electrical equivalent circuit of a piezo-electric crystal is shown in Fig. 6.12. The inductance L represents the inertia of the mass of the crystal when it is vibrating; the capacitance C_1 represents the reciprocal of the stiffness of the plate; and the resistance R represents the frictional losses of the vibrating plate. The capacitance C_2 is the actual capacitance of the crystal. The $L-C_1-R$ branch of the equivalent circuit is series resonant at a frequency f_s when $\omega_s L = 1/\omega_s C_1$. Hence

Fig. 6.12 Electrical equivalent circuit of a piezo-electric crystal.

$$f_s = \frac{1}{2\pi\sqrt{(LC_1)}} \text{ Hz.} \tag{6.21}$$

At higher frequencies $\omega L > 1/\omega C_1$ and the effective reactance of the series branch is inductive. At some frequency f_p the effective inductive reactance of the series branch has the same magnitude as the reactance of the shunt capacitor C_2. At this frequency f_p, parallel resonance occurs.

$$\omega_p C_2 = \frac{1}{\omega_p L - 1/\omega_p C_1} \qquad \omega_p C_2(\omega_p L - 1/\omega_p C_1) = 1$$

$$\omega_p^2 LC_2 - C_2/C_1 = 1 \qquad \omega_p^2 = \frac{1}{LC_2} + \frac{1}{LC_1}$$

$$\omega_p^2 = \frac{1}{LC_1}(1 + C_1/C_2) = \omega_s^2(1 + C_1/C_2)$$

$$f_p = f_s\sqrt{(1 + C_1/C_2)} \simeq f_s(1 + C_1/2C_2). \tag{6.22}$$

Since $C_1 \ll C_2$ there is very little difference between the series and parallel resonant frequencies, and usually f_p is about $1.01f_s$.

The resonant frequencies of a piezo-electric crystal are *very* stable and this means that such an element can be used as the frequency-determining section of an oscillator.

Crystals manufactured to be resonant at frequencies below 1 MHz have, typically, $L = 100$ H, $C_1 = 0.4$ pF, and $R = 2$ kΩ. Higher-frequency crystals have smaller dimensions, and typical values are $L = 20$ mH, $C = 0.2$ pF, and $R = 50$ Ω. The value of the shunt capacitance is generally some $5-10$ pF at all frequencies but this value is augmented by the various stray capacitances that appear in parallel with the crystal when it is fitted into an oscillator circuit. However, even a large change in the circuit capacitance has very little effect upon the self-resonant frequencies of a crystal, as is illustrated by the following example.

Example 6.4

A piezo-electric crystal has a series capacitance C_1 of 0.04 pF and a parallel capacitance of 10 pF. The crystal is connected into an oscillator circuit and it then has a total shunt capacitance of 15 pF connected across its terminals. Determine the percentage change in (a) the series resonant frequency, (b) the parallel resonant frequency if, due to a change of transistor, the shunt capacitance is reduced from 15 pF to 12 pF.

Solution
(a) The series resonant frequency is unaffected by the value of the shunt capacitance C_2. Hence,

% change in frequency = 0%.   (*Ans.*)

(b) From equation (6.22)

$$f_{p1} = f_s\left(1 + \frac{0.04}{2 \times 25}\right) = 1.0008 f_s \qquad (6.23)$$

and after the change in transistor

$$f_{p2} = f_s\left(1 + \frac{0.04}{2 \times 22}\right) = 1.00091 f_s. \qquad (6.24)$$

Dividing equation (6.24) by equation (6.23) gives $f_{p2} = 1.00011 f_{p1}$. The percentage change in the parallel resonant frequency is

$$\frac{1.00011 f_{p1} - f_{p1}}{f_{p1}} \times 100\% = 0.011\%. \quad (Ans.)$$

The example shows that adding extra capacitance in shunt with a crystal will decrease its parallel resonant frequency; conversely, the connection of a capacitor in series with a crystal will increase the parallel resonant frequency.

Crystal Oscillator Circuits

When the power supply is first connected to a crystal oscillator circuit a voltage pulse is applied to the crystal and causes it to vibrate at its resonant frequency. An alternating voltage at the resonant frequency is then developed between the terminals of the crystal. If this voltage is applied to the base of a transistor, it will be amplified to

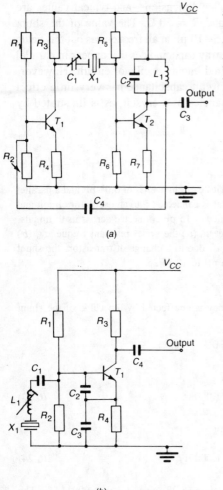

(a)

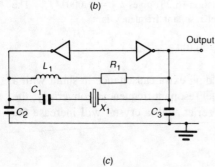

(b)

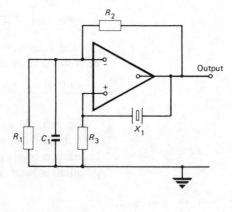

(c)

Fig. 6.13 Crystal oscillators.

appear at the collector load. If some fraction of this amplified voltage is fed back to the crystal in the correct phase, it will cause the crystal to vibrate more vigorously. A larger alternating voltage will then appear across the crystal and will be amplified and fed back to the crystal and so on. The oscillation amplitude is limited by overload when the signal driving the crystal is a square wave and the output signal is sinusoidal. The circuit will therefore oscillate at the series resonant frequency of the crystal.

A number of crystal oscillator circuits have been developed and Fig. 6.13 shows four examples. Figure 6.13(a) shows a circuit which is recommended for use at frequencies up to about 200 kHz and which uses the crystal calibrated for parallel resonance. Frequency trimming is carried out by adjustment of the capacitor C_1. The tuned circuit $L_1 C_2$ provides selectivity to ensure a sinusoidal output waveform. At frequencies of some 150 kHz to 600 kHz the circuit of Fig. 6.13(b) is often used. The inductor L_1 acts as a frequency trimmer and the crystal operates in its series-resonant mode. For higher frequencies up to about 100 MHz a similar circuit is employed, except that: (a) the output voltage is taken from the emitter resistor of T_1 and R_3 is omitted, and (b) trimming is carried out by a parallel capacitor instead of a series inductor. Figure 6.13(c) shows the circuit of a CMOS inverter crystal oscillator in which the crystal operates in the parallel-resonant mode. Finally, Fig. 6.13(d) gives a possible op-amp crystal oscillator; the crystal provides a path for the application of positive feedback to the op-amp and operates in the series mode.

7 Non-sinusoidal Waveform Generators

Waveform generators find frequent application in both analogue and digital circuitry whenever there is a need for a rectangular, a sawtooth, or (less often) a triangular waveform. Rectangular waveforms are produced by a class of oscillator known as a multivibrator. There are three main categories of multivibrator: the *monostable*, the *astable*, and the *bistable* or *flip-flop*.

An astable multivibrator operates continuously to provide a rectangular waveform with particular values of pulse repetition frequency and mark-space ratio. A monostable circuit has one stable and one unstable state; normally the circuit rests in its stable state but it can be switched into its alternative state by the application of a trigger pulse, where it will remain for a time determined by its component values. Lastly, the bistable or flip-flop circuit has two stable states and it will remain in either one until switching is initiated by a trigger pulse.

Multivibrators can be designed using bipolar transistors, FETs, operational amplifiers, logic elements such as NAND and NOR gates, and 555 timers, and are also available as integrated circuits.

A sawtooth voltage consists of a voltage that rises linearly with time, known as a ramp, until its maximum value is reached when it then falls rapidly to zero. Immediately, the voltage has fallen to zero, it begins another ramp, and so on. Sawtooth waveforms are employed whenever a voltage or current that increases linearly with time is required. More rarely a triangle-shaped waveform may be needed. Sawtooth and triangle generators can be fabricated using discrete components/devices, op-amps, and purpose-built integrated circuits.

The 555 Timer

The 555 timer was originally introduced by Signetics but it is now available from several other manufacturers. The device can be used for various timing purposes, producing accurate timing periods from a few microseconds to hundreds of seconds. It can also be connected to operate as a monostable or astable multivibrator or as a Schmitt trigger. The pin connections of the 8- and 14-pin d.i.l. packages are shown in Figs 7.1(a) and (b), respectively, while Fig. 7.1(c) shows internal block diagram of the timer.

161

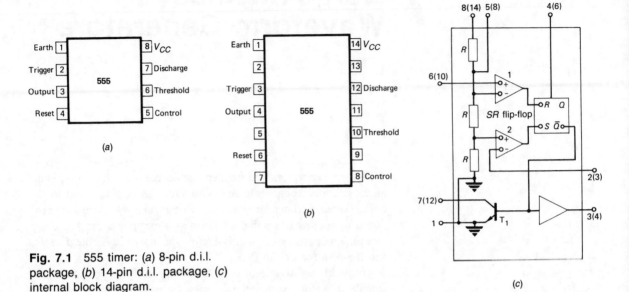

Fig. 7.1 555 timer: (*a*) 8-pin d.i.l. package, (*b*) 14-pin d.i.l. package, (*c*) internal block diagram.

Essentially the IC contains two op-amps both of which are connected as voltage comparators, one SR flip-flop, an output amplifier and a separate transistor. Also provided are three equal value resistors, labelled R, which are connected between the positive supply voltage (at pin 8 or 14) and earth. The upper comparator therefore has its inverting terminal held at $\frac{2}{3} V_{CC}$ volts, while $\frac{1}{3} V_{CC}$ appears at the inverting terminal of the lower comparator. The outputs of the two comparators are connected, respectively, to the S and R inputs of the SR flip-flop. In turn, the output of the flip-flop is connected to both the base of the transistor and the input to the output amplifier. The flip-flop can be reset by the application of a negative-going pulse to pin 4 (or 6). The control pin 5(8) can be used to vary the voltages that appear across the internal resistors. If a voltage V is applied to this pin these voltages will be equal to V and $V/2$ volts. If not used, the control terminal is connected to earth via a capacitor.

Other timer ICs that are available are the 556 dual, 558/9 quad, 7555 CMOS versions and the ZN1034; the last IC can provide timing delays of *very* long duration.

Bistable Multivibrators

A bistable multivibrator, or flip-flop, is a circuit having two stable states. The circuit will remain in one state or the other until a trigger pulse is applied to switch the circuit to its other state. It will remain in the second state until another trigger pulse is applied to switch the circuit back to its original condition. A number of variants of the flip-flop circuit are commonly employed in digital electronics; these are known as the SR, JK, T and D flip-flops, and are available in the CMOS and TTL logic families.

Op-amp Bistable Multivibrator

The bistable multivibrator can also be fabricated using an operational amplifier, the circuit being shown in Fig. 7.2. When the output of the op-amp is at its negative saturation value $V_{o(sat)}^-$, diode D_2 is ON.

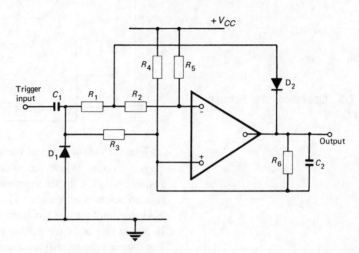

Fig. 7.2 Op-amp bistable multivibrator.

A positive trigger pulse applied to the circuit is then steered to the non-inverting terminal via R_3 and not to the inverting terminal. A positive voltage appears at the output terminals and a fraction of this voltage is fed back, via R_6 and C_2, to the non-inverting terminal as positive feedback. This causes the op-amp to rapidly drive into positive saturation. The output voltage of the circuit is then $V_{o(sat)}^+$ and this keeps the diode D_2 OFF.

The circuit will remain in this stable condition until the next positive trigger pulse is applied to the circuit. This pulse is steered, via R_2, to the inverting terminal and causes a negative voltage to appear at the output which, because of the positive feedback, causes the circuit to switch back to its negative saturation stable state. The circuit is said to toggle when a clock waveform is applied to the trigger input terminal.

Schmitt Trigger

The Schmitt trigger is a voltage comparator which is commonly employed to convert an input waveform into a rectangular waveform. The output of the circuit can only have one of two possible values. The output voltage will be high when the input voltage is greater than some threshold value and will remain at that value until such time as the input voltage falls below some other, lower, threshold value.

The difference between the two input voltages V_1 and V_2 at which switching takes place is known as the hysteresis of the circuit. Generally, this should be as small as possible, but it is typically in the region of 1 V.

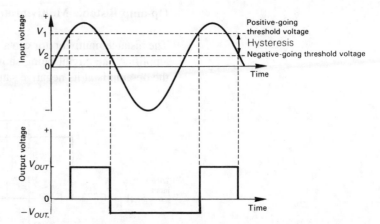

Fig. 7.3 Operation of a Schmitt trigger.

Figure 7.3 shows typical waveforms to be expected from a Schmitt trigger circuit. When the sinusoidal input voltage becomes more positive than the upper threshold voltage V_1, the circuit switches to give an output voltage of $+V_{OUT}$. The output voltage will remain at this value until the input voltage falls below the lower threshold voltage V_2 and at this point the output voltage suddenly switches to $-V_{OUT}$. The output voltage will now remain at $-V_{OUT}$ until the input voltage again exceeds the upper threshold voltage V_1. Hysteresis provides some measure of noise protection to a circuit. Once the input signal voltage has passed through a threshold voltage and the circuit has switched, the output voltage will remain constant even though the input voltage may have noise voltage superimposed upon it. Always provided, of course, that the noise voltages are not so large that they cause the input voltage to reach the other threshold value. Clearly, the larger the hysteresis, the greater the noise protection afforded.

The transfer characteristic of a Schmitt trigger circuit is shown by Fig. 7.4 and is a plot of output voltage to a base of input voltage.

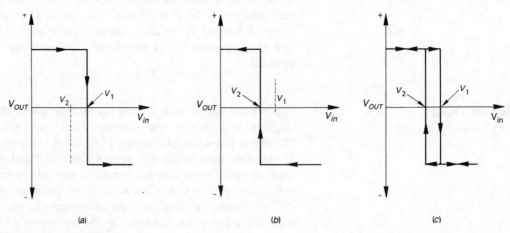

Fig. 7.4 Transfer characteristics of a Schmitt trigger.

Fig. 7.4(*a*) shows how the output voltage changes from one value to another as the input voltage is increased to the upper threshold voltage V_1 volts. Similarly, Fig. 7.4(*b*) shows how the output voltage suddenly reverts to its original value when the input voltage is reduced below the lower threshold voltage of V_2 volts. Finally, Fig. 7.4(*c*) shows the complete transfer characteristic. Note that the transfer characteristic is drawn to show the output voltage of the trigger varying between ± values, but for some circuits this would be between either ± V_{OUT} and approximately zero volts.

An op-amp can be connected to act as a Schmitt trigger, the circuit being shown by Fig. 7.5(*a*). The input voltage is applied to the inverting terminal of the op-amp and positive feedback to its non-inverting terminal.

The feedback factor is $\beta = R_1/(R_1+R_2)$ and the loop gain is βA_v.

If the output voltage V_{OUT} increases by say δV_{OUT} then the voltage fed back to the non-inverting terminal is $\beta\delta V_{OUT}$ and so V_{OUT} will be

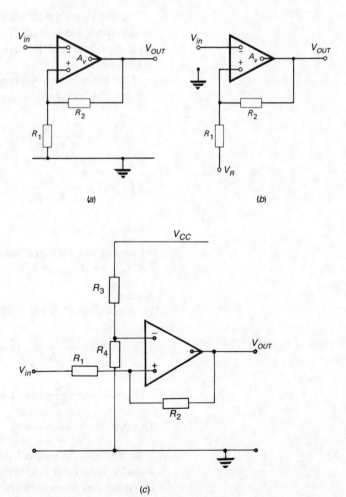

Fig. 7.5 Op-amp Schmitt triggers:
(*a*) inverting,
(*b*) inverting with reference voltage,
(*c*) non-inverting with reference voltage.

further increased by $A_v \beta \delta V_{OUT}$ and so on until positive saturation is reached. Then the voltage at the non-inverting terminal is

$$V_1 = \frac{R_1}{R_1 + R_2} \cdot V_{o(sat)}^+. \tag{7.1}$$

If now the input voltage is increased in the positive direction the output voltage will remain at $V_{o(sat)}^+$ and will be unchanged until $V_{in} = V_1$. At this threshold point, V_{in} becomes greater than V_1 and the output of the op-amp rapidly switches to its negative saturation value $V_{o(sat)}^-$ and will remain at this value as long as

$$V_{in} > V_{o(sat)}^- \cdot \frac{R_1}{R_1 + R_2}.$$

In this condition the voltage at the non-inverting terminal is

$$V_2 = V_{o(sat)}^- R_1 / (R_1 + R_2). \tag{7.2}$$

V_1 and V_2 are the threshold voltages and $(V_1 - V_2)$ is the hysteresis of the circuit and is small. The op-amp employed should have a fairly high slew rate and for most purposes the 741 is likely to prove inadequate.

If R_1 is taken to a reference voltage V_R instead of to earth (Fig. 7.5(b)), the threshold voltages equations are modified thus:

$$V_1 = V_R + \frac{R_1}{R_1 + R_2} (V_{o(sat)}^+ - V_R) \tag{7.3}$$

$$V_2 = V_R - \frac{R_1}{R_1 + R_2} (V_{o(sat)}^- + V_R). \tag{7.4}$$

Example 7.1

An op-amp Schmitt trigger has $R_1 = 560\ \Omega$, $R_2 = 12\ k\Omega$, and $V_R = 1\ V$. For $V_{o(sat)}^+ = V_{o(sat)}^- = 8\ V$, calculate the hysteresis of the circuit.

Solution
From equations (7.3) and (7.4)

$$V_1 = 1 + \frac{560}{12560} (8 - 1) = 1.312\ V$$

$$V_2 = 1 - \frac{560}{12560} (8 + 1) = 0.599\ V.$$

Therefore the hysteresis is $V_1 - V_2 = 0.713\ V$. (*Ans.*)

With the op-amp connected as shown in Fig. 7.5, the output rectangular waveform is inverted relative to the input signal. A non-inverting circuit can easily be obtained by merely applying the reference voltage to the inverting terminal and the input signal to the non-inverting terminal, Fig. 7.5(c). The threshold voltages are then

$$V_1 = V_R\left(\frac{R_1+R_2}{R_2}\right) + V_{o(sat)}^+ \frac{R_1}{R_2} \tag{7.5}$$

and

$$V_2 = V_R\left(\frac{R_1+R_2}{R_2}\right) - V_{o(sat)}^- \frac{R_1}{R_2}. \tag{7.6}$$

Integrated Schmitt Trigger Circuits

The Schmitt trigger circuit is also available as an integrated-circuit device and examples in the TTL and CMOS logic families are

> TTL   7413 dual 4-input NAND Schmitt trigger
> 7414 hex inverter Schmitt trigger
> 74132 quad 2-input NAND Schmitt trigger.

Each of these devices has a positive-going threshold voltage of 1.7 V and a negative-going threshold voltage of 0.9 V, i.e. a hysteresis of 0.8 V.

> CMOS   4093 quad 2-input NAND Schmitt trigger.

This has threshold voltages of 4 V and 6 V.
The pin connections of these devices are given in Fig. 7.6.

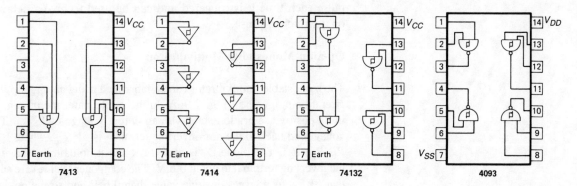

Fig. 7.6 Pin connections of four integrated Schmitt triggers.

A NAND Schmitt trigger can be used as either a NAND gate *or* as an inverting Schmitt trigger circuit, the necessary connections, referring to one 'gate' in the 4093 IC, being shown in Fig. 7.7. Fig. 7.8 shows how the 4093 can be used as a sine-to-square wave convertor or as a level detector.

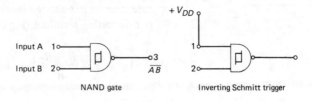

Fig. 7.7 Applications of the 4093
Schmitt trigger.

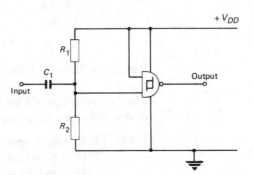

Fig. 7.8 Use of the 4093 as a level
detector.

Monostable Multivibrators

The monostable multivibrator is a circuit that has one stable state and
one unstable state. When the power supply is switched on, the circuit
will settle in its stable state and remain there until a trigger pulse is
applied to the circuit to initiate switching to the unstable state. The
circuit will then remain in this unstable state for a time that is determined by its timing components and it will then revert to its stable
condition. The monostable multivibrator generates a single output
pulse each time it is triggered and can be used to give a delayed
or a stretched pulse.

Op-amp Monostable Multivibrator

The monostable multivibrator can be fabricated using an op-amp, the
circuit being given in Fig. 7.9(*a*). In the stable state, the op-amp is
in its positive saturation state with an output voltage of $V_{o(sat)}^{+}$. The
voltage fed back to the non-inverting terminal is $\beta V_{o(sat)}^{+}$, where $\beta =
R_1/(R_1+R_2)$. The diode D_1 is turned ON and it holds the voltage at
the inverting terminal to about 0.6 V. The component values are such
that $\beta V_{o(sat)}^{+}$ is a larger positive value than 0.6 V and so the op-amp
is stable in its positive saturation state.

When a negative trigger pulse is applied to the input terminals of
the circuit, the non-inverting terminal is taken negative. The output
voltage of the op-amp goes negative and the positive feedback via
R_1 and R_2 ensures that the circuit rapidly switches back into its
negative saturation state. The output voltage of the circuit is then
$V_{o(sat)}^{-}$. This voltage turns diode D_1 OFF and this allows capacitor C_2
first to discharge and then to charge up, with the opposite polarity

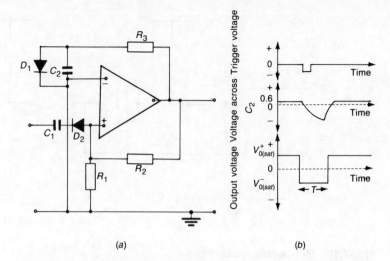

Fig. 7.9 (*a*) Op-amp monostable multivibrator; (*b*) circuit waveforms.

(*a*)

(*b*)

from previously, with a time constant of C_2R_3 seconds. Immediately the voltage across the capacitor becomes more negative than the voltage at the non-inverting terminal, the output of the circuit goes positive and, once again, the positive feedback ensures that the circuit rapidly switches back to its stable positive saturation state. Here it will remain until the next negative trigger pulse is applied to the input terminal. The circuit is in its unstable state for a time T seconds. Hence,

$$V_{o(sat)}^- - \beta V_{o(sat)}^- = (V_{o(sat)}^- - 0.6)e^{-T/C_2R_3}$$

$$\frac{V_{o(sat)}^- (1-\beta)}{V_{o(sat)}^- - 0.6} = e^{-T/C_2R_3}$$

and

$$T = C_2R_3 \log_e \left[\frac{V_{o(sat)}^- - 0.6}{V_{o(sat)}^- (1-\beta)} \right]. \tag{7.7}$$

Usually, $V_{o(sat)}^-$ is much larger than 0.6 V and then

$$T \simeq C_2R_3 \log_e [1/(1-\beta)]. \tag{7.8}$$

The waveforms in the circuit are shown by Fig. 7.9(*b*). The risetime and the falltime of the output waveform are limited by the slew rate of the op-amp.

Use of NAND/NOR Gates

Integrated-circuit NAND and NOR gates are readily available and provide a convenient and economic means of fabricating a monostable multivibrator circuit. A number of both types of gate are in both the

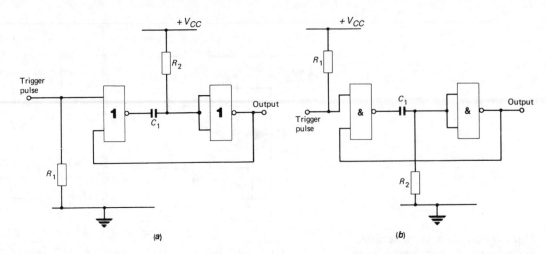

Fig. 7.10 Monostable multibrators using (a) NOR gates, (b) NAND gates.

TTL and CMOS families, providing four 2-input gates in one package. A monostable multivibrator requires the use of two NOR or NAND gates as shown in Figs. 7.10(a) and (b). In both circuits the relaxation time is determined by the time constant $C_1 R_2$.

The NOR circuit has a stable state with a low ($\simeq 0$ V) output voltage; hence in the absence of a trigger pulse the output of the left-hand gate is high. When a positive trigger pulse is applied to the circuit, the output of the input gate switches to 0 V and the change in potential is passed through C_1 to take the input of the right-hand gate to 0 V also. The output of this gate then switches high. The left-hand plate of C_1 is now at 0 V and the capacitor charges up towards $+V_{CC}$ with time constant $C_1 R_2$. When the gate input voltage reaches the threshold value, the output gate switches back to 0 V. For TTL gates the threshold voltage is 2 V and for CMOS gates it is about $V_{DD}/2$. Hence, for a TTL circuit,

$$2 = 5(1 - e^{-T/C_1 R_2}) \quad \text{or} \quad T = 0.51 \, C_1 R_2.$$

For a CMOS circuit,

$$V_{DD}/2 = V_{DD}(1 - e^{-T/C_1 R_2}) \quad \text{or} \quad T = 0.693 \, C_1 R_2.$$

The NAND monostable multivibrator works in a similar manner except that the stable state is a high output and triggering is accomplished by a negative-going voltage pulse.

Schmitt Trigger Monostable Multivibrator

A Schmitt trigger IC can be used as a monostable multivibrator, the necessary connections being shown in Fig. 7.11. In the stable condition, the n-channel MOSFET is non-conducting and C_1 is fully charged. Both inputs to the trigger are then high and so the stable state is a low output.

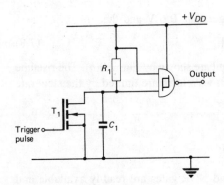

Fig. 7.11 Schmitt trigger monostable multivibrator.

When a positive voltage pulse is applied to the trigger input, T_1 is turned ON and C_1 is rapidly discharged. One input to the trigger is now low and so its output is switched to its high value. At the end of the pulse capacitor C_1 charges up via resistor R_1 and when the voltage across its terminals reaches the threshold value of the Schmitt trigger the output voltage switches back to the low value.

The expression for the voltage across C_1 is

$$v_c = V_{DD}(1 - e^{-t/R_1 C_1}).$$

The circuit will change states when this voltage reaches the threshold value V_1. This occurs after a time T, therefore,

$$V_1 = V_{DD}(1 - e^{-T/C_1 R_1}), \quad V_1/V_{DD} = 1 - e^{-T/C_1 R_1}$$

$$\frac{-T}{C_1 R_1} = \log_e\left(\frac{V_{DD} - V_1}{V_{DD}}\right), \quad T = C_1 R_1 \log_e\left(\frac{V_{DD}}{V_{DD} - V_1}\right).$$

$$(7.9)$$

Example 7.2

Calculate the width of the output pulse of a Schmitt trigger monostable multivibrator if $R_1 = 100$ kΩ, $C_1 = 0.01$ μF, $V_{DD} = 5$ V and $V_1 = 3$ V.

Solution
From equation (7.9),

$$T = 10^5 \times 10^{-8}\log_e\left[\frac{5}{5-3}\right] = 919 \ \mu s. \quad (Ans.)$$

The 555 as a Monostable Multivibrator

For the 555 timer to operate as a monostable multivibrator it must be connected as shown by Fig. 7.12(a). In the stable state, pin 2 is high, more positive than $V_{CC}/3$, so that the output of comparator 2 is low and resets the flip-flop. Then $\bar{Q}$ is at logical 1 and transistor T_1 is turned ON so that both the output terminal 3 and discharge terminal 7 are low. How low Terminal 3 goes depends upon the load current but is typically about 0.1 V. Pin 7 will be at very nearly 0 V and C_4 will be discharged. The output of comparator 1 is low since its inverting terminal is at 2 $V_{CC}/3$ volts. When a negative trigger pulse is applied to pin 2 to take the inverting terminal of comparator 2 below $V_{CC}/3$, the output of comparator 2 goes high and sets the flip-flop, so that its $\bar{Q}$ output goes low. Now transistor T_1 is turned OFF and the output voltage at pin 3 goes high. With the transistor OFF, C_4 is able to charge towards the supply voltage V_{CC} with a time constant of $C_4 R_2$ seconds. When the capacitor voltage reaches 2 $V_{CC}/3$ volts comparator 1 has its + terminal more positive than its − terminal and switches to have a high output (1.5 to 2.5 V). This

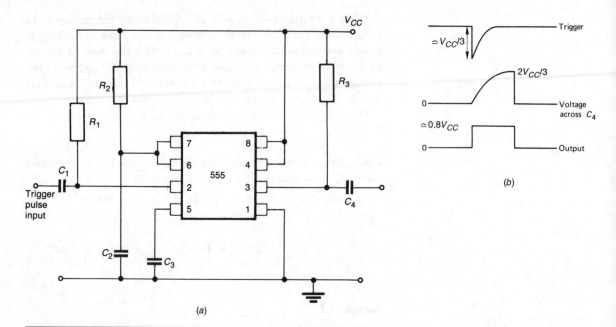

(a)

(b)

Typical values
R_1 = 2 kΩ R_2 = 100 kΩ R_3 = 1 kΩ C_1 = 0.01 μF
C_2 depends on wanted pulse width
C_3 = 0.01 μF C_4 = 0.1 μF

Fig. 7.12 555 monostable multivibrator.

resets the flip-flop so that $\bar{Q} = 1$ and the transistor T_1 again turns ON and pin 3 goes low. Thus, the output of the circuit is a positive pulse whose duration is equal to the time interval required for the capacitor to charge to 2 $V_{CC}/3$ volts. Therefore,

$$2\,V_{CC}/3 = V_{CC}(1 - e^{-t/R_2 C_4}), \quad e^{-t/R_2 C_4} = 1/3$$

$$t = C_4 R_2 \log_e 3 = 1.0986 C_4 R_2 \simeq 1.1 C_4 R_2 \text{ s.} \tag{7.10}$$

Typical waveforms are shown in Fig. 7.12(*b*).

Once the circuit has been triggered it will not respond to any other pulses that may appear at the trigger input terminal until the timing period, determined by $C_4 R_2$, has been completed. The rest terminal allows C_4 to be discharged prematurely to interrupt the timing cycle and return the output to zero. As long as pin 4 is low, $\bar{Q} = 1$ and T_1 is ON so that C_4 cannot charge. When this facility is not wanted, pin 4 must be connected to the power supply line V_{CC}. The remaining pin, 5, is the control terminal and a voltage applied to this pin will vary the timing period and hence the pulse width by modifying the d.c. voltages set up by the three resistors. When this facility is not wanted pin 5 should be connected to earth via a capacitor of about 0.01 μF.

Integrated-circuit Monostable Multivibrators

A number of monostable multivibrators are available in both the TTL and CMOS logic families. Examples of each are shown in Fig. 7.13. The *74121* is a monostable multivibrator in which the pins labelled

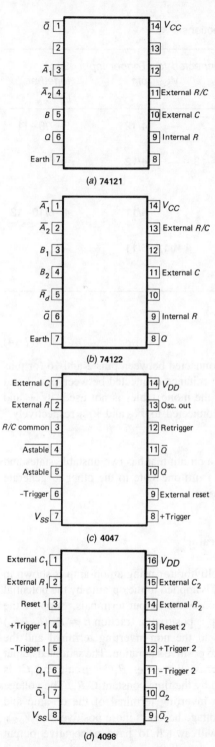

(a) 74121

(b) 74122

(c) 4047

(d) 4098

Fig. 7.13 Pin connections of four IC monostable multivibrators.

$\bar{A}_1$ and $\bar{A}_2$ are negative-edge triggered logic inputs which will trigger the monostable when either or both A_1 and A_2 go to logical 0 and pin 5 (B) is at logical 1. Pin 5 is one input to a positive Schmitt trigger (the other input being connected to the output of the gate to which $\bar{A}_1$ and $\bar{A}_2$ connect). The function of this trigger circuit is to provide level detection. Without an external capacitor or resistor the width of the output pulse is 30 ns. Pin 9 should be connected to pin 14. For other pulse widths, an external capacitor C_1 should be connected between pins 10 and 11 and an external resistor should be connected between pins 11 and 14, and pin 9 left unconnected. The output pulse width is then given by

$$T = 0.695 \, R_1 C_1. \tag{7.11}$$

The *74122* is a retriggerable monostable multivibrator which provides an output pulse width given by

$$T = 0.32 \, R_1 C_1 (1 + 0.7/R_1) \tag{7.12}$$

where C_1 is the external capacitor connected between pins 11 and 13. R_1 is the external resistor connected between pins 13 and 14. If a 10 kΩ resistance is required this can be provided internally by shorting pin 9 to pin 14 instead.

The *4047* is a CMOS device that can be used as either a monostable or as an astable multivibrator. When used as a monostable it can be operated as either a positive-edge or a negative-edge triggered device and can also be made retriggerable. The necessary pin connections for each mode of operation are given in Table 7.1.

Table 7.1 4047 pin connections

Mode of operation	Pins connected to		Trigger input pin	Output pins
	V_{DD}	V_{SS}		
+ edge triggered	4, 14	5, 6, 7, 9, 12	8	10, 11
− edge triggered	4, 8, 14	5, 7, 9, 12	6	10, 11
Retriggerable	4, 14	5, 6, 7, 9	8, 12	10, 11

For all the modes of operation the duration of the Q output pulse is given by

$$T = 2.48 \, R_1 C_1 \tag{7.13}$$

where R_1 and C_1 are the resistance and capacitance values connected between pins 2/3 and 1/3 respectively.

Finally, the *4098* is a dual monostable circuit only but it can also be operated in more than one mode of operation (see Table 7.2). The duration of the output pulse is given by

Table 7.2 4098 pin connections

Mode of operation	Pins connected to V_{DD}	V_{SS}	Trigger input pin	Short together pins
Leading-edge trigger	3/13	—	4/12	5−7/9−11
Leading-edge trigger and retriggerable	3−5/11−13	—	4/12	—
Trailing-edge trigger	3/13	—	5/11	4−6/10−12
Trailing-edge trigger and retriggerable	3/13	4/12	5/11	—

$$T = R_1 C_1 / 2 \qquad (7.14)$$

where R_1 is the resistance connected between pins 2 and 16 (or pins 14 and 16) and C_1 is the capacitance connected between pins 1 and 2 (or pins 14 and 15). If one of the monostables is not used, its + and − trigger inputs must be connected to V_{SS} and V_{DD} respectively.

Astable Multivibrators

The astable multivibrator is a circuit that has two unstable states and it will continuously change from one state to the other to generate a rectangular output waveform.

Op-amp Astable Multivibrator

The circuit of an astable multivibrator using an op-amp is shown in Fig. 7.14. Positive feedback is applied to the op-amp by the potential divider $R_2 + R_3$ connected across the output terminals. Suppose the output voltage of the op-amp is positive; a fraction $\beta = R_3/(R_2 + R_3)$ of this voltage is fed back to the non-inverting terminal and the amplifier is rapidly driven into positive saturation. The saturated output voltage $V_{o(sat)}^+$ is applied across the series $R_1 C_1$ circuit and C_1 is charged at a rate determined by the time constant $C_1 R_1$. The voltage across C_1 is applied to the inverting terminal of the op-amp and when, after a time t_1, this voltage becomes more positive than $V_{o(sat)}^+$ $R_3/(R_2 + R_3)$, the op-amp will switch to have a negative output voltage. The positive feedback will then ensure that the op-amp is rapidly driven into its negative saturation condition. Its output voltage is then $V_{o(sat)}^-$. Capacitor C_1 now starts to charge up with the opposite

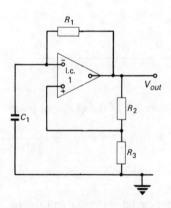

Fig. 7.14 Op-amp astable multivibrator.

polarity to before and, when after a time t_2 seconds its voltage exceeds $V_{o(sat)}^- R_3/(R_2+R_3)$, the circuit switches back to its positive saturated condition and so on.

At the moment of switching from $V_{o(sat)}^-$ to $V_{o(sat)}^+$, C_1 is charged to $\beta V_{o(sat)}^-$ volts and so the voltage at the inverting terminal suddenly increases to $-\beta V_{o(sat)}^- + V_{o(sat)}^+$ volts.

Hence at time t_1

$$\beta V_{o(sat)}^+ = V_{o(sat)}^+ - (V_{o(sat)}^+ - \beta V_{o(sat)}^-)e^{-t_1/C_1R_1} \qquad (7.15)$$

$$V_{o(sat)}^+(1-\beta) = (V_{o(sat)}^+ - \beta V_{o(sat)}^-)e^{-t_1/C_1R_1}$$

$$t_1 = C_1R_1 \, \log_e\left[\frac{V_{o(sat)}^+ - \beta V_{o(sat)}^-}{V_{o(sat)}^+(1-\beta)}\right]. \qquad (7.16)$$

Similarly,

$$t_2 = C_1R_1 \, \log_e\left[\frac{V_{o(sat)}^- - \beta V_{o(sat)}^+}{V_{o(sat)}^-(1-\beta)}\right]. \qquad (7.17)$$

The periodic time of the output waveform is $T = t_1 + t_2$. If the positive and negative saturation voltages $V_{o(sat)}^+$ and $V_{o(sat)}^-$ are equal then equation (7.15) can be written as

$$\beta V_{o(sat)} = V_{o(sat)} - V_{o(sat)}(1+\beta)e^{-t_1/C_1R_1}.$$

Then, $1-\beta = (1+\beta)e^{-t_1/C_1R_1}$ and

$$t_1 = t_2 = C_1R_1 \, \log_e\left[\frac{1+\beta}{1-\beta}\right] = C_1R_1 \, \log_e\left[\frac{1+R_3/(R_2+R_3)}{1-R_3/(R_2+R_3)}\right]$$

$$t_1 = t_2 = C_1R_1 \, \log_e[1+2R_3/R_2]. \qquad (7.18)$$

The periodic time T of the output waveform is

$$T = t_1 + t_2 = 2C_1R_1 \, \log_e(1+2R_3/R_2]. \qquad (7.19)$$

Example 7.3

The op-amp used in an astable multivibrator circuit has $V_{o(sat)}^+ = 13.5$ V and $V_{o(sat)}^- = -12$ V. Calculate the frequency of the output waveform if $R_1 = 12$ kΩ, $R_2 = 20$ kΩ, $R_3 = 51$ kΩ and $C_1 = 0.1$ μF. Calculate also the output frequency if the output saturation voltages are assumed to be equal.

Solution

$$\beta = R_3/(R_2+R_3) = 0.72$$

From equation (7.16)

$$t_1 = 1.2 \times 10^{-3} \, \log_e\left[\frac{13.5+0.72\times12}{13.5(1-0.72)}\right] = 2.12 \text{ ms}$$

and from equation (7.17)

$$t_2 = 1.2 \times 10^{-3} \log_e \left[\frac{-12 - 0.72 \times 13.5}{-12(1 - 0.72)} \right] = 2.24 \text{ ms}$$

$$T = t_1 + t_2 = 4.36 \text{ ms.}$$

Therefore the frequency $= 1/T \simeq 229$ Hz.   (*Ans.*)
If $V_{o(sat)}^+$ and $V_{o(sat)}^-$ are assumed to be equal, equation (7.19) will apply. Therefore,

$$T = 2.4 \times 10^{-3} \log_e [1 + 102/20] = 4.34 \text{ ms}$$

and $f = 1/T \simeq 230$ Hz.   (*Ans.*)

Integrated-circuit Astable Multivibrators

One of the devices quoted earlier, the CMOS *4047* (p. 173) can also be connected to operate as an astable multivibrator. Figure 7.15 shows the circuit; if gated operation is desired the gate pulse should be connected to pin 5. The frequency of the output waveform is given by equation (7.20), i.e.

$$f = 1/(4.4 \, R_1 C_1) \text{ Hz.} \tag{7.20}$$

A *4093* Schmitt trigger IC can be connected as an astable multivibrator (Fig. 7.16). Suppose the IC has just switched to have its output high. This means that input 2 must be low and hence C_1 is discharged. C_1 now commences to charge at a rate determined by the time constant $C_1 R_1$ seconds. When the voltage across the capacitor reaches the threshold voltage V_1 of the device, both of its inputs are then high and the trigger switches to its low-output state. The capacitor then discharges at the same rate until its voltage reaches the value at which the circuit again switches and so on. The periodic time of the output waveform is

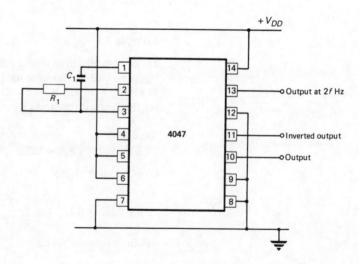

Fig. 7.15 4047 astable multivibrator.

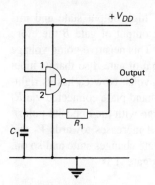

Fig. 7.16 4093 Schmitt trigger astable multibrator.

$$T = C_1 R_1 \log_e \left[\frac{V_{DD}}{V_{DD} - V_1} \right] + C_1 R_1 \log_e \left[\frac{V_{DD}}{V_{DD} - V_2} \right] \quad (7.21)$$

where V_1 and V_2 are the threshold voltages of the Schmitt trigger.

Use of NAND/NOR Gates

Both NAND and NOR gates can be used to produce an astable multivibrator circuit; the circuit is the same for each type of gate since they are employed as invertors. Figure 7.17(a) shows the circuit of a NOR astable multivibrator. If CMOS devices are used, another resistor (R_2) is often connected in series with the input to gate A to protect its input protective diodes. The output of either type of gate is high when both its inputs are low, and low when both its inputs are high.

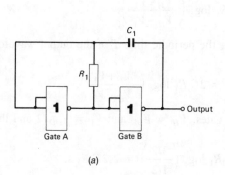

(a)

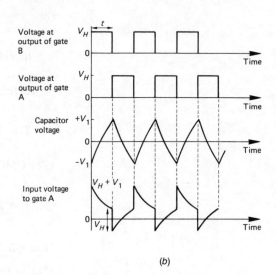

(b)

Fig. 7.17 (a) NOR gate astable multivibrator, (b) circuit waveforms.

Suppose that the circuit has *just* switched into the state with the output of gate B high at V_H volts. ($V_H = 3.6$ V for TTL and $V_{DD}/2$ for CMOS.) The input to gate B and thus the output of gate A is low, approximately 0 V, and the input to gate A must be at the *threshold voltage V_1* of the gate, i.e. the voltage at which the output of the gate changes state (2 V for TTL and $V_{DD}/2$ for CMOS). The sudden increase in gate B output voltage from 0 V to $+V_H$ volts is passed through C_1 to the input of gate A. This makes the input voltage of gate A equal to $V_1 + V_H$ volts. C_1 now has one plate at $+V_H$ volts and the other connected via R_1 to 0 V and so it commences to charge up at a rate determined by the time constant $C_1 R_1$ seconds.

As the capacitor voltage increases, the gate A input voltage falls (see Fig. 7.17(b)). When the input voltage has fallen to the threshold value V_1, the gate switches back to its high output state and this causes gate B to change state also. The output of gate B therefore falls abruptly from $+V_H$ volts to 0 volts. This negative-going voltage pulse is transferred through C_1 to the input of gate A so that its input voltage suddenly falls from V_1 to $V_1 - V_H$ volts. Now C_1 has its right-hand plate connected to 0 V and its left-hand plate connected via R_1 to V_H volts and so it commences to charge with the opposite polarity to before. The input voltage to gate A increases towards V_1 and immediately it reaches this voltage the gate changes state and so on. The expression for the input voltage to gate A is

$$v = V_H - (V_H + V_1)e^{-t/C_1 R_1}.$$

The circuit changes state when $v = V_1$, hence

$$\frac{V_H - V_1}{V_H + V_1} = e^{-t_1/C_1 R_1}$$

or

$$t_1 = C_1 R_1 \log_e \left(\frac{V_H + V_1}{V_H - V_1} \right). \tag{7.22}$$

Therefore, the periodic time T of the output waveform is

$$T = 2t_1 = 2C_1 R_1 \log_e \left(\frac{V_H + V_1}{V_H - V_1} \right). \tag{7.23}$$

For CMOS gates, $V_H = V_{DD}$ and $V_1 = V_{DD}/2$ and then

$$T = 2C_1 R_1 \log_e \left(\frac{3/2}{1/2} \right) = 2.2 C_1 R_1 \text{ s}. \tag{7.24}$$

Example 7.4

Determine the value of C_1 for a CMOS NAND gate astable multivibrator which is to operate at a frequency of 15 kHz if $R_1 = 10$ kΩ.

Solution
From equation (7.24)

$$C_1 = \frac{1}{15 \times 10^3 \times 2.2 \times 10^4} = 3 \text{ nF}. \quad (Ans.)$$

The 555 as an Astable Multivibrator

Figure 7.18 shows how the 555 IC can be connected as an astable multivibrator. The connections differ from the monostable case only in that (a) the trigger terminal 2 is now connected to the threshold terminal 6, and (b) terminals 6 and 7 are no longer connected together.

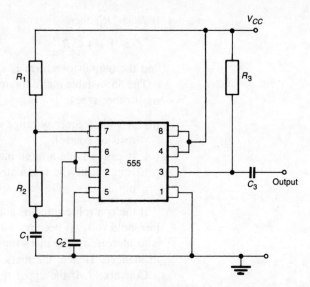

Fig. 7.18 555 astable multivibrator.

When the circuit is first switched on, C_1 charges towards V_{CC} volts with a time constant of $C_1(R_1+R_2)$. When the capacitor voltage reaches $2V_{CC}/3$ volts, the action described earlier takes place and pin 7 goes low. C_1 then starts to discharge towards 0 V into pin 7, with time constant C_1R_2. When the capacitor voltage has fallen to $V_{CC}/3$ comparator 1 switches, the flip-flop sets making $\bar{Q} = 0$ and the internal transistor T_1 turns OFF. Now C_1 charges up, via resistors R_1 and R_2, towards V_{CC} to repeat the sequence. Thus C_1 alternately charges towards V_{CC} with time constant $C_1(R_1+R_2)$ and discharges towards 0 V with time constant C_1R_2. Therefore,

$$\tfrac{2}{3}V_{CC} = V_{CC}-(V_{CC}-V_{CC}/3)e^{-t_1/C_1(R_1+R_2)}$$
$$= V_{CC}(1-\tfrac{2}{3}e^{-t_1/C_1(R_1+R_2)})$$
$$\tfrac{2}{3} = 1-\tfrac{2}{3}e^{-t_1/C_1(R_1+R_2)}$$
$$\tfrac{1}{2} = e^{-t_1/C_1(R_1+R_2)}$$

or

$$t_1 = 0.69\,C_1(R_1+R_2).$$

Also, during the discharge period,

$$\tfrac{1}{3}V_{CC} = \tfrac{2}{3}V_{CC}e^{-t_2/C_1R_2} \quad \tfrac{1}{2} = e^{-t_2/C_1R_2}$$

or

$$t_2 = 0.69\,C_1R_2.$$

The periodic time T of the output waveform is

$$T = t_1+t_2 = 0.69\,C\,(R_1+2R_2). \qquad (7.25)$$

Hence the frequency f of the output waveform is

$$f = 1/T = 1/0.69\,C_1(R_1+2R_2). \qquad (7.26)$$

If $R_2 \gg R_1$, then

$$f \simeq 1/1.4 \ C_1 R_2 \tag{7.27}$$

and the output waveform is very nearly square.

The 555 astable multivibrator has three particular advantages over most other types.

(a) A wide frequency range can be covered with a single variable resistor control.

(b) It can provide a large output current of up to 200 mA.

(c) It can easily be modulated by applying a modulating signal to pin 5.

If the control terminal is connected to $+4$ V the upper and lower threshold voltages become $+4$ V and $+2$ V respectively. The effect is to increase t_2 (the mark time) but to leave $t_2 - t_1$ (the space time) unchanged. Hence, the mark-space ratio is increased.

Conversely, if the upper threshold voltage is made smaller than $2V_{CC}/3$ volts by connecting terminal 5 to $+2$ V, then the mark-space ratio is decreased. Again the space time is not altered.

Example 7.5

A 555 timer astable multivibrator is required to operate at 2000 Hz with a mark-space ratio of 2:1. Calculate values for R_1, R_2 and C_1 if the supply voltage is 12 V and the minimum current in C_1 is to be 10 μA.

Solution

$$\frac{t_1}{t_2} = \frac{R_1 + R_2}{R_2} = 2 \quad \text{or} \quad R_1 = R_2.$$

$$T = 1/2000 = 0.5 \times 10^{-3} = 0.69 \ C_1 \times 3R_1 = 2.07 \ C_1 R_1.$$

Also,

$$\frac{V_{CC}/3}{R_1 + R_2} = \frac{4}{2R_1} \geq 10 \times 10^{-6}.$$

Hence

$$R_1 = R_2 = 200 \ \text{k}\Omega \quad (Ans.)$$

and

$$C = \frac{0.5 \times 10^{-3}}{2.07 \times 200 \times 10^3} = 1.21 \ \text{nF}. \quad (Ans.)$$

Ramp and Triangle Waveform Generators

The ramp waveform, shown in Fig. 7.19(a), is widely used as the timebase waveform in c.r.o.s and television receivers but it has many other applications as well. On the other hand, the triangle waveform (Fig. 7.19(b)) is much less often used.

For both waveforms the most important parameters are the

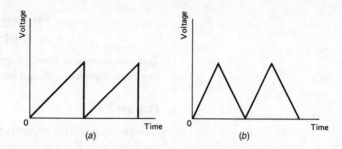

Fig. 7.19 (a) ramp, (b) triangular waveforms.

frequency and the linearity of the *ramp*. The linearity is generally expressed as the percentage of *slope error*, i.e.

$$\text{Slope error} = \frac{\text{Slope at start} - \text{Slope at end}}{\text{Slope at start}} \times 100\%. \quad (7.28)$$

The basic principle of a ramp generator is illustrated by Fig. 7.20(a). A capacitor C is charged through a resistor R from a constant voltage supply V and the capacitor voltage increases with increase in time according to

$$v_c = V(1 - e^{-t/CR}). \quad (7.29)$$

Initially the slope dv_c/dt of the waveform is fairly linear but it becomes increasingly exponential as time increases. If the switch is closed to discharge the capacitor before the markedly non-linear part of the waveform is reached, and is then opened again when the capacitor voltage reaches 0 V, a sawtooth waveform can be generated (see Fig. 7.20(b)).

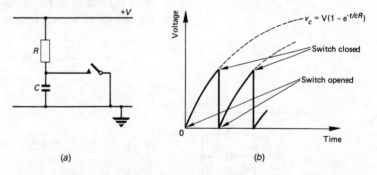

Fig. 7.20 Principle of a sawtooth waveform generator.

The departure from linearity is determined using equations (7.28) and (7.29):

$$\text{Slope} = dv_c/dt = Ve^{-t/CR}.$$

Thus the slope at the start of the ramp, $t = 0$, is V and the slope at the end of the ramp, $t = t_1$, is $Ve^{-t_1/CR}$.

$$\text{Slope error} = \frac{V - Ve^{-t_1/CR}}{V} \times 100\%$$

$$= \frac{\text{Peak voltage of sawtooth}}{\text{Supply voltage}} \times 100\%. \qquad (7.30)$$

This result means that for good linearity the peak voltage of the sawtooth waveform should be much smaller than the supply voltage.

Example 7.6

A simple ramp generator operates from a 24 V supply. Determine the percentage slope error if the peak ramp is (*a*) 10 V and (*b*) 0.1 V.

Solution
From equation (7.30),

(*a*) Slope error $= \dfrac{10}{24} \times 100\% = 41.7\%$. (*Ans.*)

(*b*) Slope error $= \dfrac{0.1}{24} \times 100\% = 0.42\%$. (*Ans.*)

Clearly, if a small percentage slope error is required, the peak ramp voltage must be small and it will be necessary to use amplification to obtain a useful output voltage.

The Miller Ramp Generator

Figure 7.21(*a*) shows the basic circuit of a Miller ramp generator. The inverting terminal of the op-amp is a virtual earth point and so an input current $I_{in} = V_{CC}/R_1$ flows through R_1 and also, since the input impedance of the op-amp is high, through C_1. Therefore,

$$V_{out} = -\frac{1}{C_1} \int I_{in}\, \mathrm{d}t = -\frac{1}{C_1} \int \frac{V_{CC}}{R_1}\, \mathrm{d}t = -\frac{1}{C_1 R_1} \int V_{CC}\, \mathrm{d}t,$$

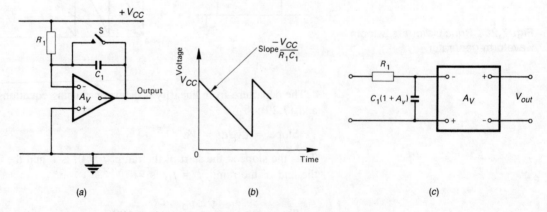

(*a*)   (*b*)   (*c*)

Fig. 7.21 Basic Miller waveform generator.

i.e. the output voltage is proportional to the time integral of the input voltage. If a constant voltage is applied, as in the figure,

$$V_{out} = -\frac{1}{C_1 R_1} \cdot V_{CC} t \qquad (7.31)$$

and is a negative-going ramp function which returns to zero when the switch S is closed (Fig. 7.21(a)). The slope error is now reduced to equation $(7.30)/A_v$. If the op-amp has any uncompensated offset errors there will then be a rate of change of output voltage even when the input voltage is zero. This will eventually cause the op-amp to saturate unless something is done to prevent it; thus the switch S must operate to discharge the capacitor before saturation has occurred.

The operation of Fig. 7.21(a) can also be explained by considering the *Miller effect* discussed earlier in this book. The equivalent circuit of Fig. 7.21(a) is shown by Fig. 7.21(c). Assuming that the input impedance of the op-amp is very high, the time constant of the circuit is $R_1 C_1 (1 + A_v)$ and hence the capacitor voltage is given by

$$v_c = V_{CC} (1 - e^{-t/R_1 C_1 (1 + A_v)}) \qquad (7.32)$$

and

$$V_{out} = A_v v_c = A_v V_{CC} (1 - e^{-t/R_1 C_1 (1 + A_v)}). \qquad (7.33)$$

The improvement in ramp linearity can be shown by expanding equations (7.32) and (7.33), using the series $e^x = 1 + x + \frac{1}{2} x^2 + \text{etc.}$ Thus for equation (7.32),

$$v_c = V_{out} = V_{CC} \left[1 - \left(1 - \frac{t}{C_1 R_1} + \frac{t^2}{2 C_1^2 R_1^2} + \cdots \right) \right]$$

$$V_{out} = V_{CC} \left[\frac{t}{C_1 R_1} - \frac{t^2}{2 C_1^2 R_1^2} \right] = \frac{V_{CC} t}{C_1 R_1} \left[1 - \frac{t}{2 C_1 R_1} \right] \qquad (7.34)$$

and for equation (7.33)

$$V_{out} = A_v V_{CC} \left[1 - \left(1 - \frac{t}{C_1 R_1 (1 + A_v)} + \frac{t^2}{2 C_1^2 R_1^2 (1 + A_v)^2} \right) \right]$$

$$= A_v V_{CC} \left[\frac{t}{C_1 R_1 (1 + A_v)} - \frac{t^2}{2 C_1^2 R_1^2 (1 + A_v)^2} \right]$$

$$= \frac{A_v V_{CC} t}{C_1 R_1 (1 + A_v)} \left[1 - \frac{t}{2 C_1 R_1 (1 + A_v)} \right]$$

$$V_{out} \simeq \frac{V_{CC} t}{C_1 R_1} \left[1 - \frac{t}{2 C_1 R_1 (1 + A_v)} \right]. \qquad (7.35)$$

The second term in equations (7.34) and (7.35) represents the departure from linearity of the ramp waveform and can be shown to be *identical* with the results obtained by applying equation (7.28).

Example 7.7

An amplifier with an inverting gain of 400 and a very high input impedance is used as a ramp generator. The d.c. supply voltage is 12 V and the resistance and capacitance used are 12 kΩ and 100 nF, respectively. Calculate the time taken to complete a 10 V ramp, and the percentage deviation from linearity at the end of the ramp.

Solution
From the first term of equation (7.35),

$$10 = \frac{12t}{100 \times 10^{-9} \times 12 \times 10^3}, \quad t = 1 \text{ ms.} \quad (Ans.)$$

From the second term, the percentage deviation from linearity in the ramp is

$$\frac{1 \times 10^{-3}}{2 \times 100 \times 10^{-9} \times 12 \times 10^3 \times 401} \times 100\% = 0.105\%. \quad (Ans.)$$

It is clear from the first term of equation (7.35) that the Miller integrator approximates to a constant current generator. The first term is $V_{CC}t/C_1R_1$ and, since V_{CC}/R_1 is the constant current, this can be written as It/C_1. Note that although a high gain is needed the actual *value* of the gain does not matter.

The switch S shown in the basic circuit Fig. 7.21(a) is normally a bipolar or a field effect transistor that is turned ON and OFF to open or close the switch.

An example is shown by Fig. 7.22. As long as the output ramp voltage is less than the reference voltage $V_{REF} = V_{CC}R_4/(R_3+R_4)$ the

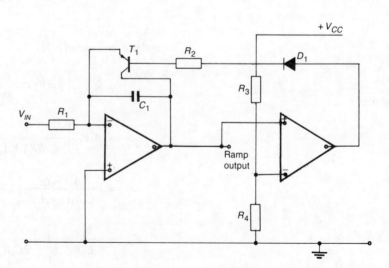

Fig. 7.22 Miller ramp generator.

output of the voltage comparator will be negative. The diode D_1 is then non-conducting and so, therefore, is transistor T_1. This means that the switch S is effectively open.

Immediately the ramp voltage becomes more positive than the reference voltage the comparator switches; its output positive voltage turns both D_1 and T_1 ON so that the switch S is effectively closed. Capacitor C_1 now discharges towards zero volts and the output voltage rapidly falls to zero. The cycle now begins once again to generate a ramp waveform at the output terminals of the circuit. The periodic time of this waveform is $T =$ (peak voltage)/(rate of change of voltage) $= V_{REF}/(V_{IN}/R_1C_1)$ and the frequency of the waveform is

$$f = \frac{V_{IN}}{V_{REF}R_1C_1}. \tag{7.36}$$

Example 7.8

Design a ramp generator to produce a peak output voltage of 5 V at a frequency of 1000 Hz. Use a supply voltage of ± 15 V.

Solution

The reference voltage should be equal to 5 V. Hence $15 R_4/(R_3+R_4) = 5$ or $R_3 = 2 R_4$. Suitable values would be $R_4 = 10$ kΩ and $R_3 = 20$ kΩ. The required periodic time is $1/1000 = 1$ ms, therefore $1 \times 10^{-3} = 5 R_1C_1/V_{IN}$. If the product R_1C_1 is made equal to 1×10^{-3} then $V_{IN} = 5$ V. Suitable values for R_1 and C_1 would then be 10 kΩ and 0.1 μF.

Example 7.9

Determine the effect on the output waveform of the circuit designed in example 7.8 if the input voltage were to be reduced to 2 V.

Solution

From equation (7.36), $f = 2/(5 \times 10^{-3}) = 400$ Hz.   (*Ans.*)
The amplitude of the ramp will be unchanged at 5 V.

The Bootstrap Ramp Generator

The circuit of an op-amp bootstrap sawtooth generator is shown in Fig. 7.23.

The voltage at the non-inverting terminal is very nearly equal to the voltage at the inverting terminal and so both ends of R_1 are at very nearly the same potential. The a.c. current which flows in R_1 must therefore be very small — regardless of the actual value of R_1 — and so the effective a.c. resistance of R_1 is very high. This bootstrap principle is used in the bootstrap ramp generator which possesses the advantage of producing a ramp voltage which starts from 0 V. For the voltage across C_1 and hence the output voltage to rise linearly it must be charged by a constant current. This can be achieved

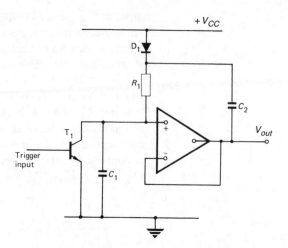

Fig. 7.23 Op-amp bootstrap ramp generator.

if a *constant* voltage can be maintained across resistor R_1. Then the constant charging current is equal to (voltage across R_1)/R_1.

Suppose that a positive trigger voltage has been applied to the circuit to turn T_1 ON and so discharge capacitor C_1. The voltage at the +terminal is then 0 V and the output voltage is also 0 V. Capacitor C_2 is charged to V_{CC} volts minus the small voltage drop across the diode D_1.

When the trigger terminal is taken to 0 V, T_1 is turned OFF and C_1 is able to commence charging towards V_{CC} volts. Because the op-amp is connected as a voltage follower its output voltage *follows* the voltage across C_1 and this charge is transferred via C_2 to the junction of D_1 and R_1. D_1 turns OFF and the potential difference across R_1 is maintained at V_{CC} volts. R_1 acts like a constant current generator to supply a constant current V_{CC}/R_1 to capacitor C_1. Then

$$V_{CC}/R_1 = C_1 \frac{dv_{c1}}{dt} = C_1 \frac{dV_{out}}{dt}$$

$$\frac{dV_{out}}{dt} = \frac{V_{CC}}{C_1 R_1} \text{ V/s.} \tag{7.37}$$

Thus the output voltage rises from zero at the constant rate of $V_{CC}/C_1 R_1$ volts/s. The ramp will not be exactly linear because C_2 is not of infinite capacitance and will therefore discharge slightly during the sweep. The maximum amplitude of the ramp is limited to V_{CC} volts. At the end of the sweep the trigger voltage turns T_1 ON again to discharge C_1. Recovery to the starting state of the ramp occupies the time needed for the voltage across C_2 to return to V_{CC} volts and this is determined by the time constant $C_2 r_{diode}$.

Example 7.10

A bootstrap ramp generator has the following component values: $R_2 = 82$ kΩ, $C_1 = 0.01$ μF and $C_2 = 10$ μF. The supply voltage is 12 V and the trigger pulse has a duration of 60 μs. Calculate (*a*) the amplitude of the ramp, (*b*) the frequency of the ramp waveform.

Solution

(*a*)   From equation (7.37)

$$V_{out} = \frac{12 \times 60 \times 10^{-6}}{82 \times 10^3 \times 10^{-8}} = 0.88 \text{ V.}   (Ans.)$$

(*b*)   The ramp occurs in 60 μs, hence it has a rate of change of 0.88 V/60 μs or 14.67 kV/s. The frequency of the ramp is

$$f = 1/T = 10^6/60 = 16.67 \text{ kHz.}   (Ans.)$$

Triangular Waveform Generation

A triangular waveform is the time integral of a square waveform and so the easiest way of generating a triangular waveform is to use the method shown in Fig. 7.24. An op-amp version of this is given in Fig. 7.25.

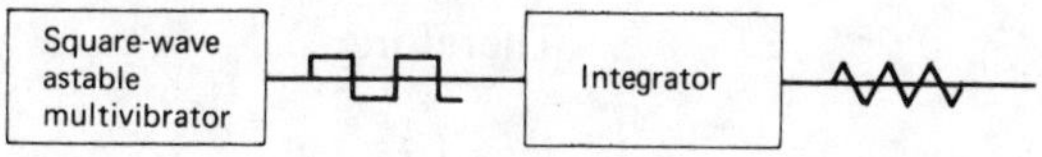

Fig. 7.24   Method of producing a triangular waveform.

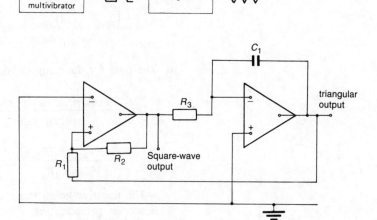

Fig. 7.25   Op-amp triangular waveform generator.

The first op-amp is connected as a Schmitt trigger whose output is switched as the ramp voltage applied to its non-inverting terminal passes through one or other of its two threshold voltages. The second op-amp is connected as an integrator which integrates the square waveform output of the first op-amp.

Suppose that the Schmitt trigger output is at its positive saturation value $V_{o(sat)}^+$, then the output of the integrator is a falling ramp waveform V_{ramp}^-. Then the voltage V_+ at the non-inverting input to the trigger is

$$V_+ = \frac{V^+_{o(sat)}R_1}{R_1+R_2} + \frac{V^-_{ramp}R_2}{R_1+R_2}.$$

The trigger switches when $V_+ = 0$ and so when

$$V^-_{ramp} = \frac{-V^+_{o(sat)}R_1}{R_2}.$$

Similarly, on the rising half of the triangular output waveform

$$V^+_{ramp} = \frac{V^-_{o(sat)}R_1}{R_2}$$

and the peak-peak triangular voltage is

$$V^+_{ramp} - V^-_{ramp} = 2\,V_{o(sat)}R_1/R_2$$

assuming equal saturation voltages.

The current charging C_1 is

$$I = \frac{V_{o(sat)}}{R_3} = \frac{C_1 dv_c}{dt} = -\frac{C_1 dV_{ramp}}{dt}.$$

Therefore,

$$\frac{+V_{o(sat)}}{C_1 R_3} = \frac{dV_{ramp}}{dt} = \text{rate at which the ramp voltage increases (or decreases).}$$

The time for the ramp to be completed is

$$t_1 = \frac{\text{Peak-peak ramp voltage}}{\text{Ramp rate}}$$

$$= \frac{2\,V_{o(sat)}R_1/R_2}{V_{o(sat)}/C_1 R_3} = \frac{2\,R_1 R_3 C_1}{R_2}.$$

The negative-going ramp voltage occupies an equal time t_2 and hence the periodic time $T = t_1 + t_2$ of the triangular waveform is

$$T = 4R_1 R_3 C_1/R_2.$$

Therefore frequency $= 1/T = R_2/4R_1 R_3 C_1$ Hz. (7.38)

The 555 as a Ramp Generator

The circuit of a 555 ramp generator is shown by Fig. 7.26. The operation of the circuit is similar to that of the monostable multivibrator except that capacitor C_3 is now charged by the *constant* current produced by the transistor T_2. This means that the capacitor voltage rises linearly with time until pin 7 goes low and flyback is initiated. C_3

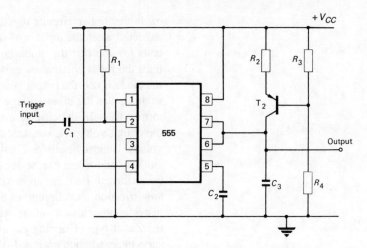

Fig. 7.26 555 ramp generator.

still charges from 0 V to 2 $V_{CC}/3$ and the time for a ramp is $\frac{2}{3}V_{CC}/i$ where i is the constant current provided by T_2.

Clamping and Clipping

Clipping circuits are often used to limit the positive and/or negative peak(s) of a signal to some required value.

Such circuits provide a means of obtaining a rectangular waveform from a sinusoidal signal. A variety of clipping circuits exist but here only diode circuits will be discussed. Figure 7.27 shows four examples

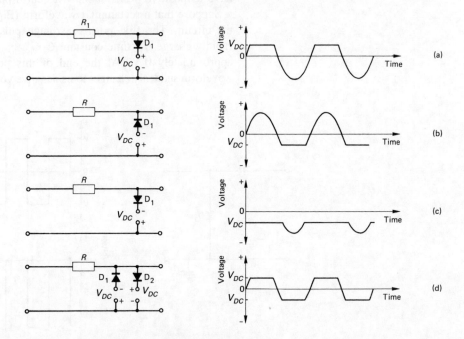

Fig. 7.27 Diode clipping circuits.

of diode clipper circuits together with the output waveforms to be expected when the input is of sinusoidal waveform. Referring to circuits (a) and (b): the diode D_1 is held OFF by the d.c. bias voltage until the input voltage exceeds V_{DC} when the diode turns ON. Once the diode is ON, the output voltage remains constant at V_{DC} volts until such time as the input voltage falls below V_{DC} and allows the diode to turn OFF. In the case of circuit (c), the diode is always ON and the output voltage is constant at $-V_{DC}$ volts except when the input voltage is more negative. Finally, circuit (d) is a combination of circuits (a) and (b); it can be seen that the input sinusoidal voltage has been turned into an approximation of a square wave. Further amplification and clipping of this waveform would produce a much better approach to a square waveshape. Sometimes the diode is of the Zener type. Clipping circuits are used to remove noise voltages superimposed upon a wanted signal and for the improvement of pulse waveforms.

Whenever a waveform is passed through a capacitor or a transformer, its d.c. component is lost. For many applications the loss of the d.c. component cannot be tolerated and in such cases some means of establishing a d.c. level is necessary. A clamping circuit is shown in Fig. 7.28(a). When a positive voltage is applied to the circuit, diode D_1 is turned ON and connects a very low-resistance path across the output terminals. Conversely, when a negative input voltage is applied, D_1 is OFF and the input voltage appears at the output terminals. This means that any input alternating waveform will have its positive peaks *clamped* to earth potential.

Suppose that a rectangular waveform (Fig. 7.28(b)) is applied to the circuit. The first positive voltage pulse will turn D_1 ON and rapidly charge C_1 (time constant $C_1 r_{diode(on)}$). The output voltage is approximately 0 V. At the end of this positive pulse, the input waveform suddenly changes to its negative voltage, a change of -2 V

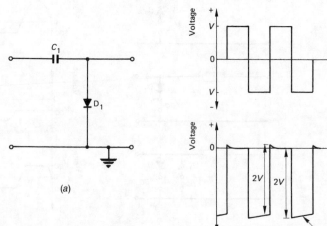

Fig. 7.28 Direct current clamping.

volts. The voltage across a capacitor cannot change instantaneously and so the right-hand terminal of C_1 must also experience a voltage change of -2 V volts. Capacitor C_1 now commences to discharge through the diode but, since the *off-resistance* of D_1 is high, the discharge time constant is long. Capacitor C_1 does not therefore discharge very much during this time period and the output voltage of the circuit remains at -2 V. When the next positive pulse arrives, the input voltage rises positively by 2 V and so the output voltage becomes 0 V and so on (see Fig. 7.28(c)).

A reversal in the polarity of the diode D_1 will result in the negative peaks of the waveform being clamped to 0 V.

Example 7.11

Draw the output waveform of the circuit shown in Fig. 7.29(a) if the diode has an ON resistance of 100 Ω and an OFF resistance of 200 kΩ.

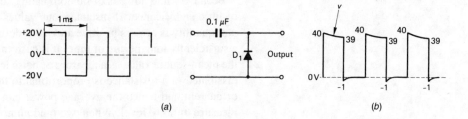

Fig. 7.29
        (a)   (b)

Solution

When the input is positive the circuit time constant is

$$0.1 \times 10^{-6} \times 2 \times 10^{5} = 20 \times 10^{-3} \text{ s.}$$

When the input is negative the time constant is

$$0.1 \times 10^{-6} \times 100 = 10 \text{ μs.}$$

Hence the output waveform is as shown in Fig. 7.29(b).

8 Noise

The output waveform from any electronic circuit will always contain some unwanted voltages and currents in addition to the wanted signal. These unwanted components are known as *noise* or *interference* and can originate from a variety of different sources. Some of these noise sources are situated within the circuit itself, while others will already be present when the signal first appears at the input terminals. This means that the amplification process will *always* degrade the output signal-to-noise ratio. Noise having a constant energy per unit bandwidth over a particular frequency band is said to be *white noise*.

Because white noise is of random nature, instantaneous values are of little significance. If instantaneous values are squared, the resultant quantity is always positive and has a definite average value over a sufficiently long period of time. It is customary therefore to employ the mean-square value as a measure of noise level. The average power dissipated in a resistance is proportional to the mean-square voltage or current and hence the average power can also be employed as a measure of noise level. When two random noise voltages or currents are applied in series, the resultant voltage is the arithmetic sum, instant by instant, of the individual mean-square voltages. The sum waveform fluctuates randomly in the same manner as the waveforms of the component waves. The mean-square voltage of the sum waveform is equal to the sum of the mean-square voltages of the component noise voltages; similarly the power of the sum waveform is equal to the sum of the powers of the component waves.

The noise power per unit bandwidth is known as the *power density spectrum* (p.d.s.). The p.d.s. of a noise source tells how the noise power is distributed over the frequency spectrum. The p.d.s. of white noise is a horizontal line as shown by Fig. 8.1 and it has an equal amount of power in each hertz of the bandwidth. For white noise

$$\text{p.d.s.} = \frac{\text{Total noise power in band between } f_1 \text{ and } f_2}{\text{bandwidth } (f_2 - f_1) \text{ (Hz)}} \text{ W.}$$

(8.1)

Example 8.1

If the total noise power in the band $1-2$ MHz is 80 μW determine the p.d.s. if the noise can be assumed to be white.

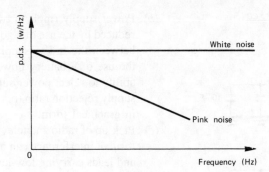

Fig. 8.1 Power density spectrum of white and pink noise.

Solution
From equation (8.1).

$$p.d.s. = \frac{80 \times 10^{-6}}{1 \times 10^{-6}} = 8 \times 10^{-11} \text{ W/Hz.} \quad (Ans.)$$

The term *pink noise* is given to noise that has a p.d.s. which is inversely proportional to frequency (see Fig. 8.1). Pink noise has an equal amount of power in each decade of bandwidth, i.e. there is the same noise power in the frequency bands 0.1−1 Hz, 1−10 Hz, and 10−100 Hz.

Sources of Noise

Noise can be random or repetitive (impulsive) in its nature, be internally or externally generated and be of varying frequency and bandwidth. Whatever its nature the effects of noise can always be reduced by good design.

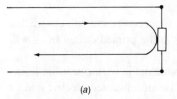

(a)

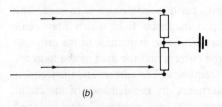

(b)

Fig. 8.2 (a) Symmetrical (b) non-symmetrical interference currents.

External Noise Sources

There are a large number of possible sources of external noise and interference that are located in different parts of the frequency spectrum. Interference may be either symmetrical or non-symmetrical. Symmetrical interference currents flow in opposite directions in the two conductors of the signal path, or the live and neutral conductors of a mains supply, see Fig. 8.2(a). Non-symmetrical interference currents flow in the same direction in both conductors and return via earth, Fig. 8.2(a).

Among the more important sources of interference to an electronic circuit are the following.

(a) Repetitive interference at the mains frequency caused by unwanted couplings with nearby power lines and/or transformers. These couplings may be either electric or magnetic and can generally be reduced by careful positioning of the power leads, screening the mains transformer, and by the use of a battery supply.

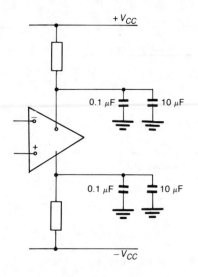

Fig. 8.3 Decoupling power supply lines to reduce interference.

(*b*) Power-supply ripple at twice the mains frequency. This can be reduced by the use of a good filter circuit, decoupling capacitors between the $\pm V_{CC}$ supply lines (see Fig. 8.3) and, of course, the use of a battery power supply. Op-amps have an inherent ability to reject power supply ripple; it is known as the power-supply rejection ratio (p.s.r.r.) and is often quoted in data sheets in graphical form.

(*c*) Pick-up of radio and television (and perhaps radar) signals. This form of interference can be minimized by screening components and leads carrying low-level signals and by keeping all leads as short as possible. This will also reduce any interference that is generated by the screened equipment itself.

(*d*) Whenever an electric current is switched on and off spikes of voltage are generated because of the unavoidable inductance and capacitance of the circuit. A voltage spike may be transient or a large number may occur continuously. Transient spikes occur when a circuit is switched on and off either mechanically or electronically; examples of mechanical switching are electric light switches, and electric motors. Continuous interference spectrums are produced by such circuits as thyristor speed control and switched-mode power supplies. Switching circuits generate interference components at the switching frequency and at the harmonics of the switching frequency, while phase-control circuits generate a continuous interference spectrum.

The interference may be either electro-magnetic (e.m.i.) or it may be radio-frequency (r.f.i.). The former is of lower frequency, 10 kHz to 30 MHz, but is usually of high voltage and it is conducted over the mains wiring for a considerable distance. The latter is of high frequency, 30−300 MHz, and it is both conducted over, and radiated from, the mains wiring which acts like a rather inefficient aerial.

(*e*) Mechanical vibrations of an electronic equipment will ensure that any loose connections or cracks in the printed circuit track will cause short-duration breaks.

(*f*) Crosstalk can occur between conductors in wiring or cable forms and between p.c.b. tracks because of unwanted electric and/or magnetic couplings. Figure 8.4 shows a p.c.b. with four ICs mounted on it. Each IC has its pin 5 connected to earth. If current flows from pin 8 of IC 1 to pin 4 of IC 4 through a track on the p.c.b. and thence to earth, a small loop is formed. This current loop will set up a magnetic field which may cause interference with the other ICs. The magnitude of the crosstalk is proportional to both the current and the area of the loop and also to the square of the frequency. The current and the frequency are, presumably, determined by the demands of the circuit design, and so careful attention to the component layout and track geography is essential to minimize loop area. Basically, this means keeping each track as close as possible to its associated earth return.

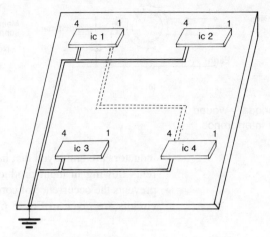

Fig. 8.4 Interference in p.c.b.

Another problem may occur because the earth-return system on the p.c.b. is not of zero resistance; this may result in some parts of the earth-return plane having a common-mode voltage above 0 V. If a conductor is connected to such parts it will have this potential applied to it and will pass an unwanted current, the conductor may then act as an aerial. This effect can be minimized by the use of capacitors connecting the affected parts of the p.c.b. to earth.

Reduction of Interference

The reduction of interference to an acceptable level can be achieved by the use of filters and screens, by careful layout of the components and the routing of cables, tracks etc., and by careful earthing. These steps should include both the source of the interference and each electronic equipment.

A capacitor can be connected between the live and the neutral wires of the mains supply to suppress symmetrical interference, and between the two wires and earth to suppress non-symmetrical interference. The connections are shown by Fig. 8.5. The arrangement will operate both to prevent interference being fed *from* the supply and also to stop interference generated within the equipment being passed on to the mains conductors.

Improved suppression can be obtained by the addition of inductance in series with one, or both, conductors. Often toroidally wound

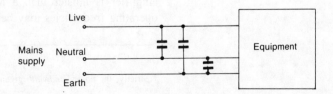

Fig. 8.5 Suppression of interference.

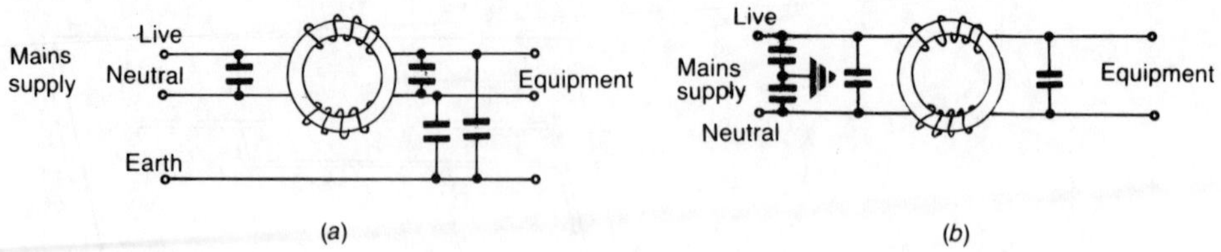

Fig. 8.6 Use of toriodally-wound inductors to reduce interference.

inductors are employed that have balanced windings so that the currents flowing in them produce opposing, equal, core fluxes. This prevents the occurrence of core saturation. Two possible suppression circuits are given in Fig. 8.6.

Internal Noise Sources

Internal noise sources are random in their nature and so their magnitudes can be mathematically predicted.

Thermal Agitation Noise

The electrons in a conductor are in a state of thermal agitation and this results in minute currents flowing in the conductor which vary continuously and randomly in both magnitude and direction. Because of this a randomly varying voltage is developed across the conductor, and this unwanted voltage is known as thermal agitation noise or *resistance noise*.

The r.m.s. value of this voltage is given by equation (8.2), i.e.

$$V_n = \sqrt{[4kTBR]} \tag{8.2}$$

where: k = Boltzmann's constant = 1.38×10^{-23} J/K;

T = *absolute* temperature of conductor in K. Note that K = °C+273;

B = bandwidth (Hz) over which the noise is measured or of the circuit at whose output the noise appears, whichever is the smaller;† and

R = resistance, or real part of the impedance, of the circuit in ohms.

It is the *bandwidth* and not the frequency of operation that is important with regard to thermal agitation noise. Thus a wideband amplifier is noisier than a narrowband amplifier whatever their operating frequencies may be. Thermal agitation noise is white.

---

† Strictly, the *noise bandwidth* given by $B_n = \displaystyle\int_0^\infty \frac{G(f)df}{G_{max}}$ should be used.

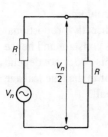

Fig. 8.7 Available noise power.

Example 8.2

Calculate the noise voltage produced in a 100 kΩ resistor in a 4 MHz bandwidth at a temperature of 20 °C.

Solution
From equation (8.2)

$$V_n = \sqrt{[4 \times 1.38 \times 10^{-23} \times 293 \times 10^6 \times 10^5]} = 80.4 \ \mu V. (Ans.)$$

The thermal noise e.m.f. may be regarded as acting in series with the resistance producing it. Maximum power transfer from one resistor to another occurs when their resistances are equal. Consider Fig. 8.7, which shows a resistance connected across another resistance of the same value. The *maximum* noise power P_a delivered to the load resistance is

$$P_a = \frac{(V_n/2)^2}{R} = \frac{4kTBR}{4R} = kTB \quad \text{watts.} \tag{8.3}$$

The maximum or *available* noise power that can be delivered by a resistance is proportional to *both* temperature *and* bandwidth.

If the two resistors are at equal temperatures they will each supply kTB watts noise power to the other and zero net transfer of energy will take place. If the resistors are at different temperatures there *is* a net power transfer that is proportional to the difference between their absolute temperatures.

Example 8.3

Calculate the available noise power from a resistance at 17 °C in a 1 MHz bandwidth.

Solution

$$P_a = 1.38 \times 10^{-23} \times 290 \times 10^6 = 4 \times 10^{-15} \text{ W.} \quad (Ans.)$$

Generally, resistors are at room temperature, nominally 290 K, and this is denoted by the symbol T_0.

Noise in Semiconductors

Thermal Agitation Noise
Thermal agitation noise voltages are generated in the resistances of the base, collector, and emitter regions of a bipolar transistor and in the channel resistance of a FET.

Shot Noise
Shot noise in a transistor is caused by the random arrival and departure of charge carriers across a p-n junction. Since there are two p-n junctions in a bipolar transistor there are two sources of shot noise.

Shot noise will also be present in a semiconductor diode. The magnitude of each source of shot noise depends upon both the *d.c.* current *I* flowing across the junction and the bandwidth *B*.

$$\text{Shot noise} = \sqrt{[2eIB]} \tag{8.4}$$

where e = electronic charge = 1.6×10^{-19} coulombs.

Shot noise has a mean square value which is directly proportional to the bandwidth and this means that the p.d.s. is constant and hence shot noise is white. Note particularly that shot noise does not depend upon the temperature. Some shot noise does appear in the gate leakage current of a jFET but this is normally very small.

Example 8.4

Calculate the shot noise produced by a semiconductor diode when it is passing a current of 2 mA and the signal bandwidth is 100 kHz.

Solution

$$\text{Shot noise} = \sqrt{(2 \times 1.6 \times 10^{-19} \times 2 \times 10^{-3} \times 10^5)} = 8 \text{ nA.} (Ans.)$$

Flicker Noise

Flicker noise is also often known as $1/f$ noise and is caused by fluctuations in the conductivity of the semiconductor material. Flicker noise is most significant at low frequencies and is generally negligible at frequencies in excess of 10 kHz or so. Flicker noise is pink.

Burst or Popcorn Noise

Burst noise is another source of low-frequency noise that is exhibited by some devices and it is given this name because it consists of bursts of noise that characteristically sound like 'popping'. This source of noise is most prevalent in op-amps.

The noise sources within a circuit are not correlated with one another and are best represented by referring all noise to the input terminals. The noise generated within any circuit can be represented by equivalent noise current and voltage generators. These generators are connected in the input of the circuit as shown by Fig. 8.8(*b*) and this allows the circuit to be considered as noiseless.

The values of the two generators can be fairly easily determined since

V_n = equivalent input r.m.s. noise voltage in nV/$\sqrt{\text{Hz}}$ at a specified frequency, or in μV in a given frequency band, that would produce the same noise output as the circuit with its input terminals short-circuited, and

I_n = equivalent input r.m.s. noise current in pA/$\sqrt{\text{Hz}}$ or nA in a given bandwidth that would produce the same noise output as the circuit with its input terminals open-circuited. I_n is zero for a FET input.

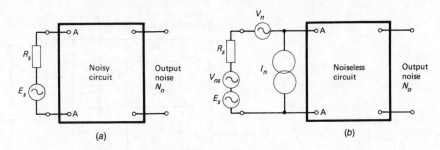

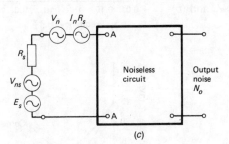

Fig. 8.8 (a) Amplifier fed by a source of e.m.f. E_s and resistance R_s, (b) noise equivalent circuit of (a), (c) Thevenin version of (b).

Equivalent Noise Resistance

It is sometimes more convenient to work in terms of *equivalent noise resistances*. These are the (imaginary) resistances in which thermal agitation would generate the noise voltage V_n or the noise current I_n. Therefore

$$V_n^2 = 4kTBR_{nv} \quad \text{or} \quad R_{nv} = \frac{V_n^2}{4kTB} \tag{8.5}$$

$$I_n^2 = 4kTBG_{ni} = \frac{4kTB}{R_{ni}} \quad \text{or} \quad R_{ni} = \frac{4KTB}{I_n^2}. \tag{8.6}$$

For a bipolar transistor

$$R_{nv} \simeq r_{bb'} + 1/2\, g_m \quad \text{and} \quad R_{ni} \simeq 2\, h_{fe}/g_m. \tag{8.7}$$

For a FET

$$R_{nv} \simeq 0.7/g_m \quad \text{and} \quad R_{ni} \simeq 0. \tag{8.8}$$

The total noise generated within an op-amp is quoted by the manufacturer in the form of graphs showing how both V_n and I_n vary with frequency. Figure 8.9 shows two typical graphs: the two characteristics are both the summation of a line of slope $-1/2$, that represents the low-frequency flicker noise, and a line of zero slope, that represents white noise. If both these lines are projected, as shown by Fig. 8.10, they will intersect at a frequency, known as the *corner frequency*, at which the flicker and white noise powers are equal. Typical values for the corner frequencies are: (a) voltage, 1–20 Hz for bipolar transistor input op-amps and 100–500 Hz for FET input types; and (b) current, bipolar input op-amps have 10–1000 Hz.

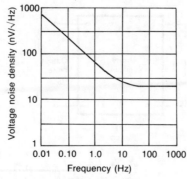

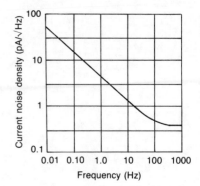

Fig. 8.9 Variation of (a) V_n and (b) I_n with frequency.

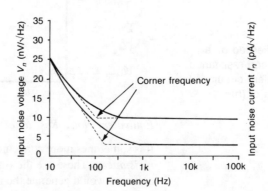

Fig. 8.10 Determination of the corner frequency.

Some manufacturers show the current noise as being due to two separate generators, one connected to each of the op-amp inputs.

Noise Calculations

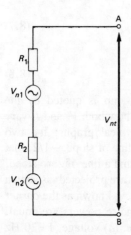

Fig. 8.11 Noise generators in series.

The internal noise sources of a transistor and of an op-amp either are the result of thermal agitation or are semiconductor noise. In either case they can be represented by a current or a voltage generator, or an equivalent noise resistance. When calculating the noise performance of a circuit, the total noise arising from several sources is determined using the *mean square value* of each noise generator. Consider Fig. 8.11 which shows two resistors R_1 and R_2 connected in series. The r.m.s. noise voltages produced by the two resistors are

$$V_{n1} = \sqrt{[4kTBR_1]} \quad \text{and} \quad V_{n2} = \sqrt{[4kTBR_2]}.$$

The total mean-square noise voltage appearing between the terminals A and B is

$$V_{nt}^2 = V_{n1}^2 + V_{n2}^2 = 4kTB(R_1+R_2)$$

and clearly this is the same result as would be obtained if the thermal agitation noise in the combined resistance R_1+R_2 had been calculated. This would not be true however if the two resistors were at different temperatures.

A similar result is obtained for two resistors connected in parallel; the total mean square noise voltage is then

$$V_{nt}^2 = 4kTBR_1R_2/(R_1+R_2).$$

Applying Thevenin's theorem to Fig. 8.8(b) gives the noise equivalent circuit shown in Fig. 8.8(c). From this the total mean-square noise voltage V_{nt}^2 at the input terminals of the 'noiseless' circuit is

$$V_{nt}^2 = V_{ns}^2 + V_n^2 + I_n^2 R_s^2.\dagger \tag{8.9}$$

Using R_{nv} and R_{ni} equation (8.9) can be written as

$$V_{nt}^2 = 4kTB(R_s+R_{nv}+R_s^2/R_{ni}). \tag{8.10}$$

Example 8.5

Calculate the total equivalent noise voltage per unit bandwidth for an op-amp operating at 1 kHz from a 2 kΩ source. The op-amp has the noise-density/frequency characteristic given by Fig. 8.12. $T = 300$ K.

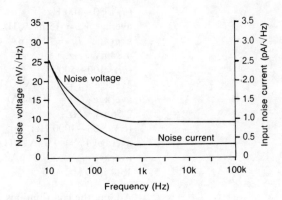

Fig. 8.12

Solution

Noise generated by source

$$= \sqrt{(4 \times 1.38 \times 10^{-23} \times 300 \times 2000)} = 5.75 \text{ nV}/\sqrt{\text{Hz}}.$$

From Fig. 8.12,

$$V_n = 9.5 \text{ nV}/\sqrt{\text{Hz}}$$
$$I_n = 0.3 \text{ pA}/\sqrt{\text{Hz}},$$

so, $I_n R_s = 0.6$ nV/$\sqrt{\text{Hz}}$.

Therefore, total equivalent noise voltage

$$= \sqrt{(5.75^2 + 9.5^2 + 0.6^2)}$$
$$= 11.1 \text{ nV}/\sqrt{\text{Hz}}. \quad (Ans.)$$

---

† Really a term representing any correlation between the two generators should be included but since it is usually small it will be neglected.

Example 8.6

Calculate the total noise voltage of the op-amp in Example 8.5 over the frequency band (*a*) 1 kHz–10 kHz, and (*b*) 10–1000 Hz.

Solution

(*a*) Since the graphs are flat over the band 1 kHz to 10 kHz the total noise voltage is equal to the product of noise (per unit bandwidth) and $\sqrt{}$bandwidth. Therefore,

$$V_n = 11.1 \times \sqrt{9000} = 1.05 \ \mu V. \quad (Ans.)$$

(*b*) The bandwidth must be split into a number of sections and in each of these the average values of V_n and I_n estimated. These average values should then be squared, multiplied by the source resistance in the case of I_n, multiplied by the section bandwidth, and then the square root of the sum of each section should be calculated.

(*i*)   10–50 Hz:
average V_n = 21 nV/$\sqrt{}$Hz, V_n^2 = 441 nV^2/Hz,
average I_n = 2 pA/$\sqrt{}$Hz, $I_n R_s$ = 4 nV/$\sqrt{}$Hz, $I_n^2 R_s^2$ = 16 nV^2/Hz.

(*ii*)   50–100 Hz:
average V_n = 15 nV/$\sqrt{}$Hz, V_n^2 = 225 nV^2/Hz,
average I_n = 1.1 pA/$\sqrt{}$Hz, $I_n R_s$ = 2.2 nV/$\sqrt{}$Hz, $I_n^2 R_s^2$ = 4.84 nV^2/Hz.

(*iii*)   100–500 Hz:
average V_n = 11.5 nV/$\sqrt{}$HZ, V_n^2 = 132 nV^2/Hz,
average I_n = 0.8 pA/$\sqrt{}$Hz, $I_n R_s$ = 1.6 nV/$\sqrt{}$Hz, $I_n^2 R_s^2$ = 2.56 nV^2/Hz.

(*iv*)   500–1000 Hz:
average V_n = 9.6 nV/$\sqrt{}$Hz, V_n^2 = 92 nV^2/Hz,
average I_n = 0.6 pA/$\sqrt{}$Hz, $I_n R_s$ = 1.2 nV/$\sqrt{}$Hz, $I_n^2 R_s^2$ = 1.44 nV^2/Hz.

Adding the contributions of V_n^2 and $I_n^2 R_n^2$ in each section of the bandwidth and multiplying by the bandwidth of that section gives the following.

(*i*)   10–50 Hz: $(441 + 16) \times 40 = 18\ 280$ nV^2.
(*ii*)   50–100 Hz: $(225 + 4.84) \times 50 = 11\ 492$ nV^2.
(*iii*)   100–500 Hz: $(132 + 2.56) \times 400 = 53\ 824$ nV^2.
(*iv*)   500–1000 Hz: $(92 + 1.44) \times 500 = 46\ 720$ nV^2.
Total noise voltage = $\sqrt{(18\ 280 + 11\ 492 + 53\ 824 + 46\ 720) \times 10^{-18}}$
                = 361 nV.   (*Ans.*)

When an op-amp is to be used in an application demanding low-noise operation three factors should be considered carefully. These are: (*a*) the chosen op-amp should have both a low corner frequency and low white noise, (*b*) the source resistance should be as low as possible, and (*c*) the bandwidth of the circuit should be no wider than is necessary to pass the signal.

Noise Factor (or Figure)

The noise factor, or noise figure, F of a circuit is defined as

$$F = \frac{\text{total noise power at output}}{\text{that part of the above which is due to the thermal noise at the source}} \quad (8.11)$$

$$= \frac{N_0}{GkT_0B} \quad (8.12)$$

The source is supposed to be at an absolute temperature of 290 K and the circuit must be linear. The terms noise figure and noise factor are both commonly used but in this book figure will be preferred.

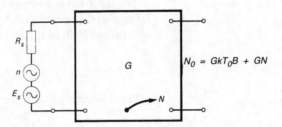

Fig. 8.13 Circuit for the definition of noise figure.

Figure 8.13 shows a circuit of *available power* gain† G, that has internal noise power (referred to the input terminals) N, fed by a source of e.m.f. E_s and thermal noise voltage n. From equation (8.12),

$$F = \frac{N_0}{GkT_0B} = \frac{1/kT_0B}{G/N_0} = \frac{P_{in}/kT_0B}{GP_{in}/N_0}$$

where N_0 is the output noise power, P_{in} is the input signal power, and G is the available power gain of the circuit. Therefore

$$F = \frac{\text{Input signal-to-noise ratio}}{\text{Output signal-to-noise ratio}}. \quad (8.13)$$

Noise figure is a measure of the degradation of the input signal-to-noise ratio caused by a circuit but it is *not* a measure of the output signal-to-noise ratio itself.

Example 8.7

If the signal-to-noise ratio at the input to an amplifier having a noise figure of 10 is 40 dB, calculate the output signal-to-noise ratio.

---

† The term *available power gain* implies that both the input and output terminals of the circuit are matched to their respective source and load impedances.

Solution

Noise figure is essentially a power ratio and hence $F = 10$ corresponds to $F = 10$ dB. From equation (8.13)

Output signal-to-noise ratio $= 40 - 10 = 30$ dB.   (*Ans.*)

A circuit with a noise figure of 3 dB or less would be regarded as a low-noise circuit while a noise figure of 10 dB would be average.

The noise figure of a circuit can be determined in terms of either the noise generators V_n and I_n, or the equivalent noise resistances R_{nv} and R_{ni}, using Fig. 8.8(*c*). The *noiseless circuit* will amplify both signal and noise to the same extent so that the output signal-to-noise ratio of the circuit will be the same as that at the input terminals AA of the amplifier.

The input signal-to-noise ratio is $E_s^2/V_{ns}^2 = E_s^2/4kTBR_s$. The signal voltage at the terminals AA is E_s and the total noise at AA is given by equation (8.9) or (8.10). Therefore, the output signal-to-noise ratio is

$$E_s^2/(V_{ns}^2 + V_n^2 + I_n^2 R_s^2) = E_s^2/[4kTB(R_s + R_{nv} + R_s^2/R_{ni})].$$

From equation (8.13) the noise figure of the circuit is

$$F = 1 + \frac{V_n^2 + I_n^2 R_s^2}{V_{ns}^2} = 1 + \frac{R_{nv}}{R_s} + \frac{R_s}{R_{ni}}. \tag{8.14}$$

For a FET input circuit

$$F = 1 + \frac{V_n^2}{V_{ns}^2} = 1 + \frac{R_{nv}}{R_s}. \tag{8.15}$$

Equation (8.14) shows that the noise figure obtained for a particular circuit depends not only upon the noise sources within the circuit but also on the source resistance R_s. Clearly, F will be very large *either* if R_s is large *or* if R_s is small and this is an indication that there must be some optimum value for R_s at which the minimum possible value for the noise figure is obtained. The optimum value for R_s can be determined by differentiating F with respect to R_s and equating the result to zero. Thus,

$$\frac{dF}{dR_s} = \frac{-R_{nv}}{R_s^2} + \frac{1}{R_{ni}} = 0.$$

Hence $R_{s(optimum)} = \sqrt{[R_{nv}R_{ni}]} = V_n/I_n.$ \tag{8.16}

Substituting equation (8.16) into equation (8.14) gives

$$F_{min} = 1 + \frac{R_{nv}}{\sqrt{[R_{nv}R_{ni}]}} + \frac{\sqrt{[R_{nv}R_{ni}]}}{R_{ni}}$$

$$= 1 + 2 \sqrt{\frac{R_{nv}}{R_{ni}}} = 1 + \frac{2 V_n I_n}{4 kT_0 B}. \tag{8.17}$$

Example 8.8

An op-amp has $V_n = 550$ nV and $I_n = 55$ pA over a 3 kHz bandwidth. Calculate the optimum source resistance and the minimum noise figure for the op-amp.

Solution
From equation (8.16).

$$R_{s(optimum)} = (550 \times 10^{-9})/(55 \times 10^{-12}) = 10 \text{ k}\Omega \quad (Ans.)$$

and from equation (8.17)

$$F_{min} = 1 + \frac{2 \times 550 \times 10^{-9} \times 55 \times 10^{-12}}{4 \times 1.38 \times 10^{-23} \times 290 \times 3000} = 2.26 = 3.54 \text{ dB}.$$

$$(Ans.)$$

The choice of source resistance to give the minimum possible noise figure does *not* give the lowest possible output noise power. With $R_s = 10$ kΩ, as in example 8.8, $I_n R_s = 55 \times 10^{-12} \times 10^4 = 550$ nV, but if $R_s = 1$ kΩ, $I_n R_s$ would only be equal to 55 nV. This means that an existing value of source resistance should not be increased, by the addition of series resistance, to make it equal to the optimum value. If, however, an input transformer is employed its turns ratio could be chosen to give the optimum source resistance.

Noise Output of a Circuit

The available noise power into a circuit is kT_0B watts and the noise factor can be written as

$$F = \frac{P_{in}/kT_0B}{GP_{in}/N_0}$$

$$F = N_0/GkT_0B \quad \text{and} \quad N_0 = FGkT_0B. \tag{8.18}$$

GkT_0B is the amplified input noise power and so the noise power N_0' due to sources within the amplifier appearing at the output is given by

$$N_0' = (F-1)GkT_0B. \tag{8.19}$$

The internal noise power referred to the input is $N_i = (F-1)kT_0B$.

Variation of Noise Figure with Frequency

Since the gain, or loss, of a circuit and the noise generated per unit bandwidth are often a function of frequency, noise figure may be frequency dependent. Thus a distinction between single-frequency noise figure and integrated or full-band noise figure must be made. If the bandwidth B is narrow enough for variations in gain and generated noise to be ignored, the *spot noise figure* is obtained. The spot noise figure can be measured at a number of different points in the overall bandwidth of the amplifier and then plotted against frequency. If the bandwidth is wide, the average noise figure is obtained. If the spot noise figure is constant over a specified bandwidth, it will be equal to the average noise figure in that bandwidth.

Overall Noise Figure of Several Circuits in Cascade

Figure 8.14 shows two circuits connected in cascade. The circuits have noise figures F_1 and F_2 and power gains G_1 and G_2. It is assumed that the bandwidths of the two circuits are equal to one another and are also equal to the overall bandwidth of the cascade.

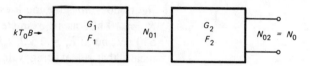

Fig. 8.14 Noise figures in cascade.

The available output noise power is

$$N_0 = F_{12}G_1G_2kT_0B \tag{8.20}$$

where F_{12} is the overall noise figure. For the first circuit

$$N_{01} = F_1G_1kT_0B$$

and for the second circuit

$$N_{02} = N_0 = G_2F_1G_1kT_0B + (F_2-1)G_2kT_0B. \tag{8.21}$$

Comparing equations (8.20) and (8.21)

$$F_{12} = F_1 + \frac{F_2-1}{G_1}. \tag{8.22}$$

The derivation has assumed that the bandwidths of the two stages are equal to one another, or that the second stage has a narrower bandwidth. Should the bandwidth B_2 of the second stage be wider than the bandwidth B_1 of the first stage then equation (8.22) must be written as

$$F_{12} = F_1 + \frac{(F_2-1)B_2}{GB_1}. \tag{8.23}$$

Similarly it can be shown that for n circuits connected in cascade

$$F_{1n} = F_1 + \frac{F_2-1}{G_1} + \frac{F_3-1}{G_1G_2} + \ldots \frac{F_n-1}{G_1G_2 \ldots G_{n-1}}. \tag{8.24}$$

Equations (8.23) and (8.24) show that the effect of noise sources in stages other than the first is reduced, unless the first stage introduces loss, because of the division by the gains of all the preceding circuits. If the first stage introduces loss, the noise introduced by the second stage will be accentuated; this means that any loss in the first circuit must always be minimized.

Example 8.9

An amplifier of power gain 20 dB and noise figure 5 dB is connected in cascade with a 6 dB attenuator to provide an overall gain of 14 dB. Determine the overall noise figure obtained with each of the two possible cascase arrangements.

Solution

Equation (8.22) has been derived assuming the power gains and noise figures of the amplifier to be quoted as power ratios. Hence the first step is to convert the given dB values into the corresponding power ratios. Thus

5 dB = 3.16:1 power ratio, 6 dB = 3.98:1 power ratio,
14 dB = 25.12:1 power ratio, and 20 dB = 100:1 power ratio.

The noise figure of an attenuator is equal to its attenuation, i.e. $F = 6$ dB.

(*a*) With the amplifier followed by the attenuator

$$F_{12} = 3.16 + \frac{(3.98-1)}{100} = 3.19 = 5 \text{ dB}. \quad (Ans.)$$

(*b*) With the attenuator followed by the amplifier

$$F_{12} = 3.98 + \frac{(3.16-1)}{1/3.98} = 12.58 = 11 \text{ dB}. \quad (Ans.)$$

Effective Noise Figure of a Circuit Fed by a Source at $T_s \neq T_0$

The noise figure of a circuit is defined with reference to a temperature of 290 K. If the source is not at 290 K but at some other temperature T_s the degradation of the input signal-to-noise ratio is different from that indicated by the noise figure. For example, if the source is at a temperature below 290 K the degradation is greater but if the source temperature is above 290 K the degradation in signal-to-noise ratio is reduced. This can be taken account of by the concept of the *effective noise figure* F_{eff}.

From equation (8.12) the noise figure of a circuit referred to temperature $T_0 = 290$ K is

$$F = N_0/GkT_0B$$

and hence the available output noise power due to the network alone is

$$(F-1)GkT_0B \text{ watts.}$$

If, now, a source at temperature T_s is connected to the input terminals of the circuit, the available output noise power will be

$$N_0 = GkT_sB + (F-1)GkT_0B.$$

The effective noise figure F_{eff} is

$$F_{eff} = N_0/GkT_sB$$

and substituting for N_0

$$F_{eff} = \frac{GkT_sB+(F-1)GkT_0B}{GkT_sB} = 1 + \frac{T_0}{T_s}(F-1). \tag{8.25}$$

Noise Temperature

The available noise output power N_0 generated within a circuit may be considered as originating from its output resistance which is at a temperature T_n where

$$T_n = \frac{N_0}{GkB}. \tag{8.26}$$

T_n is the *noise temperature* of the circuit and it is related to the standard temperature T_0 by $T_n = tT_0$ where t may be greater than, or smaller than, unity.

There is a simple relationship between the noise figure and the noise temperature of a circuit and it can be obtained in the following manner. The noise produced within a circuit is

$$GkT_nB = (F-1)GkT_0B.$$

Therefore $T_n = (F-1)T_0$ or $t = F-1$. $\tag{8.27}$

The *total noise* output from a circuit (see equation 8.18) can now be written as

$$N_0 = GkB(T_s + T_n) \tag{8.28}$$

where T_s is the temperature of the source which may, of course, be equal to T_0, see Fig. 8.15.

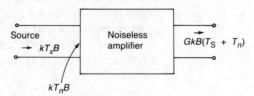

Fig. 8.15 Total output noise.

Example 8.10

An amplifier has a bandwidth of 100 kHz, a power gain of 30 dB, and a noise factor of 6 dB, and it is fed by a source at temperature T_0. Calculate the output noise power using (*a*) equation (8.18), (*b*) using equation (8.28).

Solution

(a) $N_0 = 4 \times 10^3 \times 1.38 \times 10^{-23} \times 290 \times 10^5 = 1.6 \times 10^{-12}$ W. (*Ans.*)

(b) $t = F - 1 = 3$ and $T = tT_0 = 3 \times 290$. Therefore

$$N_0 = 10^3 \times 1.38 \times 10^{-23} \times 10^5 (3 \times 290 + 290) = 1.6 \times 10^{-12} \text{ W}.$$

(*Ans.*)

This example makes it clear that either noise figure or noise temperature can be used to determine output power or output signal-to-noise ratio.

Noise Temperatures in Cascade

If 1 is subtracted from both sides of equation (8.24),

$$F_{1n} - 1 = F_1 - 1 + \frac{F_2 - 1}{G_1} + \frac{F_3 - 1}{G_1 G_2} + \text{etc.}$$

$$t_{1n} = t_1 + \frac{t_2}{G_1} + \frac{t_3}{G_1 G_2} + \text{etc.}$$

$$T_{1n} = T_1 + \frac{T_2}{G_1} + \frac{T_3}{G_1 G_2} + \text{etc.} \tag{8.29}$$

Measurement of Noise Figure and Noise Temperature

The measurement of the noise figure of a circuit is usually carried out using some kind of noise generator. For frequencies from 0 Hz up to some hundreds of megahertz, the noise generator is usually a semiconductor diode. The shot noise produced by a diode is given by equation (8.4), i.e.

$$i_n = \sqrt{[2eIB]}$$

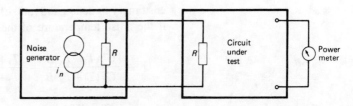

Fig. 8.16 Measurement of noise figure.

Figure 8.16 shows the arrangement used to measure the noise figure of a circuit. The r.m.s. noise current produced by the noise generator can be set to any desired value by varying the current passed by the diode. The noise generator has an output resistance R, which is matched to the input resistance of the circuit under test. The total noise output power developed by the noise generator is

$$N_0 = KT_0 B + (i_n/2)^2 R$$

or $N_0 = kT_0B + 2eIBR/4 = kT_0B + \frac{1}{2}eIBR$.

This is also the input noise to the circuit under test. With the noise generator switched off, the reading of the power meter, or true r.m.s. ammeter, at the output of the circuit is noted. This noise power is equal to $N_0 = FGkT_0B$ watts. The noise generator is then switched on and its noise output is increased until the reading of the power meter is doubled. Then

$$2FGkT_0B = FGkT_0B + GeIBR/2$$

or $FGkT_0B = GeIBR/2$ Therefore,

$$F = \frac{eIR}{2kT_0} = 20IR. \tag{8.30}$$

Noise figure measurements can be employed to determine the values of the equivalent noise generators R_{nv} and R_{ni}. If the noise figure is measured with a very low value of R_s the result will enable R_{nv} to be calculated; a second measurement of F with R_s at a very high value will then give R_{ni}.

The standard method of measuring the noise temperature of a circuit is to apply two noise sources of known temperatures T_1 and T_2 in turn to the circuit and then to obtain the ratio Y of the output powers thus obtained. Two separate standard noise sources are used only if precise measurements are being carried out; commonly a single noise generator is used and this is switched between two known temperatures.

The block diagram of the arrangement used to measure noise temperature is shown by Fig. 8.17. With the noise source at effective noise temperature T_1 connected to the circuit the reading of the output meter is noted. The circuit is then connected to the other noise generator at temperature T_2 (or the temperature of the sole generator is altered to T_2) and the new output meter reading is noted. It is customary to denote the ratio of the two meter readings as Y. Hence, if the noise temperature of the circuit is T_A,

$$Y = \frac{Gk(T_1+T_A)B}{Gk(T_2+T_A)B} = \frac{T_1+T_A}{T_2+T_A}$$

$$T_A = \frac{T_1 - YT_2}{Y-1}. \tag{8.31}$$

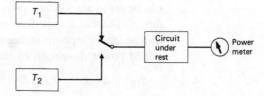

Figure 8.17 Measurement of noise temperature.

9 The Phase-locked Loop

A phase-locked loop is a closed-loop feedback system in which the fed-back signal is the phase difference, or *phase error*, between two voltages, rather than the more usual current or voltage. The basic action of the loop is to synchronize, or *lock*, the frequency of a *voltage-controlled oscillator* (v.c.o.) to the frequency of the incoming signal.

The concept of the phase-locked loop (p.l.l.) has been known for many years but the complexity of the circuit meant that it was rarely economically practical. Now a number of integrated circuit versions of the p.l.l. are offered by different manufacturers at low cost and this has made the p.l.l. technique popular for various signal-detection and processing applications.

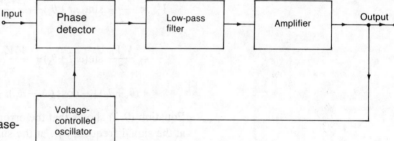

Fig. 9.1 Block diagram of a phase-locked loop.

The basic block diagram of a p.l.l. is shown by Fig. 9.1; it can be seen to consist of a phase detector, a low-pass filter, an amplifier and a v.c.o. Most integrated p.l.l.s require only a few external components to complete the system; usually these are components which allow the user to set the free-running frequency of the v.c.o. and the passband of the filter.

Components of the Phase-locked Loop

The Phase Detector

A phase detector is a circuit which compares the signal waveforms applied to its two input terminals and develops an output voltage that is proportional to the *phase difference* between the two waveforms. The output voltage is known as the *error voltage*, V_e. If the two input signals are at different frequencies the error voltage will contain components at a number of different frequencies; amongst which will be a component at a frequency equal to the difference between the input frequencies. This is known as the difference-frequency component. Should the two input signals be at the *same* frequency their difference frequency will be zero and a d.c. error voltage will be obtained.

The action of the phase detector is to *multiply* its two input signals together. In a p.l.l. one of the input signals to the phase detector is the signal to be processed and the other signal is provided by the v.c.o. The output voltage of the v.c.o. is usually a square waveform varying between $+V_0$ volts and approximately zero volts; hence its instantaneous voltage v_0 is given by equation (9.1)

$$v_0 = V_0 \left[\tfrac{1}{2} + \frac{2}{\pi} \sin(\omega_0 t + \theta_0) + \frac{2}{3\pi} \sin(3\omega_0 t + \theta_0') + \ldots \right].$$

$$(9.1)$$

If the input signal voltage is of sinusoidal waveform, i.e. $v = V_s \sin(\omega_s t + \theta_s)$ volts, the output voltage V_e of the phase detector will be

$$V_e = \frac{V_s V_0}{2} \sin(\omega_s t + \theta_s) + \frac{2 V_s V_0}{\pi} \sin(\omega_s t + \theta_s) \sin(\omega_0 t + \theta_0) + \ldots$$

$$= \frac{V_s V_0}{2} \sin(\omega_s t + \theta_s) + \frac{V_s V_0}{\pi} [\cos\{(\omega_s + \omega_0)t$$

$$+ (\theta_s + \theta_0)\} - \cos\{(\omega_s - \omega_0)t + (\theta_s - \theta_0)\}]. \qquad (9.2)$$

Equation (9.2) shows that the error voltage V_e contains components at the signal frequency f_s, at the sum and the difference frequencies $f_s \pm f_0$, and at the sum and the difference of the input signal frequency and the odd harmonics of the v.c.o. waveform. The low-pass filter which follows the phase detector will not pass any components that are of higher frequency than $f_s - f_0$. The output of the filter is hence

$$V_d = \frac{V_s V_0}{\pi} \cos[(\omega_s - \omega_0)t + (\theta_s - \theta_0)]. \qquad (9.3)$$

The amplitude of V_d is a function of both the signal and the v.c.o. voltages and of the phase difference between them. Usually, the input signal voltage is made large enough to ensure that signal limiting takes place so that V_d does not depend upon the signal voltage. The amplitude of the error voltage is then equal to the *gain factor* K_D of the phase detector, i.e.

$$V_d = K_D \cos[(\omega_s - \omega_0)t + (\theta_s - \theta_0)]. \qquad (9.4)$$

When the signal and the v.c.o. frequencies are the same, i.e. ω_s

$= \omega_0$, $V_d = K_D \cos(\theta_s - \theta_0) = K_D\cos\theta_e$, where θ_e is the *static phase error*. It is usual to express the phase error in terms of the departure from a quadrature (90°) relationship because it is when $\theta_e = 90°$ that the error voltage is zero, therefore,

$$V_d = K_D \sin\theta_e. \tag{9.5}$$

If the static phase error is small, $\sin\theta_e \simeq \theta_e$ (in radians) and

$$V_d = K_D\theta_e. \tag{9.6}$$

The operation of a phase detector is illustrated by the waveforms given in Fig. 9.2 in which it is assumed that the input signal is of

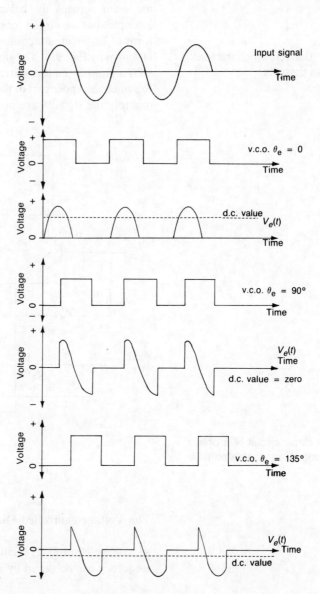

Fig. 9.2 Phase detector waveforms.

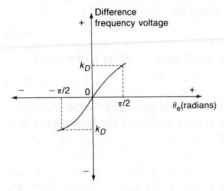

Fig. 9.3 Conversion characteristic of a phase detector.

sinusoidal waveform and of the same frequency as the v.c.o. When the two signals are in phase with one another the error voltage V_e is a positive d.c. voltage. When the signal voltage leads the v.c.o. voltage by 90° the error voltage is zero so this is the phase relationship that is taken as the reference. If the phase error becomes greater than 90° the error voltage will become a negative d.c. voltage as has been shown for $\theta_e = 135°$.

Figure 9.3 shows the conversion characteristic of a phase detector; it can be seen that the characteristic is linear for small values of the static phase error, θ_e, and that operation must be restricted to the region $\theta_e = \pm \pi/2$ radians.

A variety of different circuits can be operated as a phase detector, with perhaps the balanced modulator being the best-known example, but the circuit most often employed within an IC is some form of differential amplifier. The basic circuit of such a detector is shown by Fig. 9.4. The signal is applied to a differential amplifier with a constant emitter-current source. The v.c.o. signal is applied to switch the upper part of the circuit on and off and so effectively to multiply the signal waveform by a unity square wave.

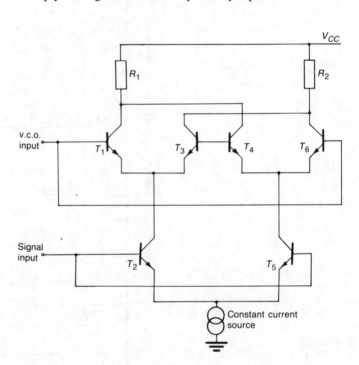

Fig. 9.4 Basic circuit of a phase detector within an integrated p.l.l.

The Voltage-controlled Oscillator

A voltage-controlled oscillator (v.c.o.) is an oscillator whose frequency can be varied by a voltage applied to its control terminal.

If the control voltage is quiescent the v.c.o. will operate at its *free-running, centre, frequency* f_c. When the control voltage has changed from its quiescent value by V_d volts the instantaneous angular velocity ω_0 of the v.c.o. output waveform is

$$\omega_0 = \omega_c + K_0 V_d. \tag{9.7}$$

K_0 is the *conversion gain* of the v.c.o. and it relates the v.c.o. frequency to the control voltage. Thus

$$K_0 = \Delta\omega_0/V_d \text{ kHz/V} \quad \text{or} \quad \text{rad/s/V}. \tag{9.8}$$

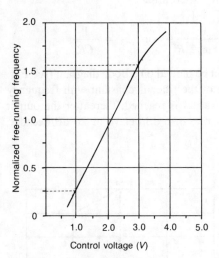

Fig. 9.5 Frequency/voltage characteristic for a v.c.o.

A typical output-frequency/control-voltage characteristic for a v.c.o. is given in Fig. 9.5. The normalized free-running frequency is given on the vertical axis since this can be set by the user to any value within a given range. The conversion gain K_0 of the v.c.o. can be obtained from Fig. 9.5. Suppose that the free-running frequency is 10 kHz and that the control voltage is varied from 1.0 V to 3 V, then $K_0 = (1.51 - 0.25) \times 10^4/2 = 6.3$ kHz/V. From equation (9.8), $\Delta\omega_0 = K_0 V_d$: integrating $\Delta\theta_0 = \int K_0 V_d \mathrm{d}t = K_0 V_d/S$. This means that the phase of the v.c.o. output is proportional to the integral of the control voltage, and that the v.c.o. acts like an integrator in the feedback loop.

The main requirements for the v.c.o. are: (*a*) a linear frequency/voltage characteristic, (*b*) free-running frequency easily adjusted to a wanted figure, and (*c*) good frequency stability. The v.c.o. is usually some form of astable multivibrator.

The Low-pass Filter

The basic function of the low-pass filter is to pass only the difference-frequency component of the error voltage output of the phase detector. The filter also improves the response of the p.l.l. to changes in the frequency of the input signal and reduces the effects of any interference and noise. A fairly simple resistance–capacitance filter is normally perfectly adequate and the two filter circuits most often employed are given in Figs. 9.6(*a*) and (*b*). The transfer function $F(S)$ of the simpler filter, Fig. 9.6(*a*), is

$$F(S) = \frac{V_{out}(S)}{V_{in}(S)} = \frac{1/SC}{R + 1/SC} = \frac{1}{1 + SCR}. \tag{9.9}$$

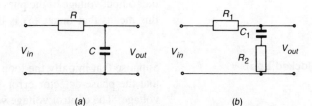

Fig. 9.6 Two p.l.l. low-pass filters.   (a)   (b)

For a sinusoidal input signal voltage $S = j\omega$ and

$$F(j\omega) = \frac{1}{1+j\omega CR}. \tag{9.10}$$

At the *cut-off frequency* f_3 of the filter the magnitude of $F(j\omega)$ is 3 dB down on its maximum value. Therefore,

$$|F(j\omega)| = \frac{1}{\sqrt{2}} = \frac{1}{\sqrt{(1+\omega_3^2 C^2 R^2)}} \quad \text{or} \quad \omega_3 = \frac{1}{CR}.$$

The response of the filter can be plotted on a Bode diagram (see Fig. 9.7(a)). The output voltage of the filter is constant with frequency at 0 dB until the breakpoint at ω_3 is reached; thereafter the output voltage falls at the constant rate of 6 dB/octave or 20 dB/decade.

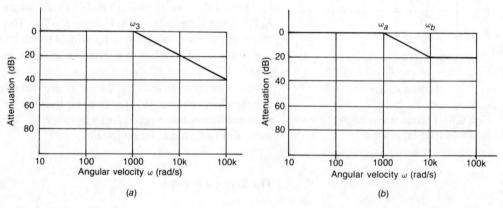

Fig. 9.7 Bode diagrams for Fig. 9.6(a) and (b).

(a)

(b)

The transfer function of the other filter (Fig. 9.6(b)), is

$$F(S) = \frac{R_2 + 1/SC}{R_1 + R_2 + 1/SC} = \frac{1+SCR_2}{1+SC(R_1+R_2)}. \tag{9.11}$$

The Bode plot for the filter is shown in Fig. 9.7(b). Above the first breakpoint, which occurs at $\omega_a = 1/C(R_1+R_2)$, the output voltage falls at 20 dB/decade until the second breakpoint, at $\omega_b = 1/CR_1$, is reached. Thereafter, the output voltage remains at the value reached at ω_b.

When the p.l.l. is in lock the difference frequency is zero and the d.c. output voltage of the phase detector is not attenuated by the filter; this means that then $F(S)$ is unity.

The Phase-locked Loop

Suppose that initially the loop is broken at the input to the v.c.o. so that the phase-detector error voltage is not fed back as the control voltage. The control voltage will be at its quiescent value so that the v.c.o. runs at its free-running frequency f_c. If there is an input signal at a frequency f_s the phase detector will multiply the v.c.o. and input signals together and generate an output error voltage V_e. Provided the two frequencies are not too far apart the difference-frequency, $f_s - f_c$, component will appear at the output of the filter. This voltage will then be amplified to become the output voltage of the system.

If the loop is now closed the amplified error voltage $A_V V_d$ is applied to the v.c.o. as a control voltage. The free-running frequency of the v.c.o. will then be modulated by the control voltage so that the difference frequency, $\Delta f = f_s - f_0$, varies with time. When the v.c.o. frequency deviates towards the signal frequency, Δf is reduced and the control voltage V_d varies more slowly with time. Conversely, if the v.c.o. frequency is shifted away from the signal frequency, Δf is increased and the variations of the control voltage are more rapid. As a result the waveform of the control voltage V_d is non-sinusoidal and has an average, or d.c., value which moves the v.c.o. frequency nearer to the signal frequency. Once the difference frequency Δf is small enough the feedback around the loop causes the v.c.o. to synchronize, or *lock*, with the incoming signal. The difference-frequency Δf is then zero, see Fig. 9.8. The number of cycles of the control voltage waveform that occur before a lock is established may be more, or less, than the number shown in the figure. Once lock has been established the frequency f_0 of the v.c.o. is equal to the frequency of the input signal but there is *always* a phase difference between the two waveforms. The static phase error $\theta_e = \theta_s - \theta_0$ must exist so that the phase detector is able to generate the error voltage needed to keep the v.c.o. at the frequency of the incoming signal. Since the difference frequency is zero hertz a d.c. control voltage is generated.

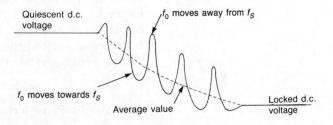

Fig. 9.8 Variation of control voltage as lock is established.

Example 9.1

A p.l.l. uses a v.c.o. which has a free-running frequency of 100 kHz and a conversion gain of 20 kHz/V. If lock has been established with a 110 kHz input signal what control voltage has been applied to the v.c.o.?

Solution

$\Delta f = 110 - 100 = 10$ kHz.

Therefore, control voltage $V_d = 10$ kHz/20 kHz/V $= 0.5$ V.   (*Ans.*)

The bandwidth over which the p.l.l. can acquire lock with an incoming signal is known as the *capture range* of the loop.

Once a p.l.l. has locked to an incoming signal it will be able to follow, or *track*, any changes in the frequency of that signal. If the input signal frequency should change there will be an instantaneous change in the phase error θ_e, with a resultant change in the error voltage output of the phase detector. This change will be in the direction required for the v.c.o. to shift to the new signal frequency. Consequently, the *average* change in the difference frequency will be zero. This means that the p.l.l. has the ability to track changes in the frequency of the input signal *once* lock has been established. The range of frequencies over which the input signal frequency can alter without the p.l.l. losing lock is known as the *lock range* of the loop. The lock range is limited by two factors: (*a*) there is a maximum value to the possible frequency deviation of the v.c.o., and (*b*) if the input signal voltage is weak the error voltage developed by the phase detector will be proportional to both signal voltage and phase error. The lock range is not affected by the low-pass filter, although the filter does affect the maximum rate at which tracking can take place.

Before lock has been established the difference-frequency output of the phase detector may be of too high a frequency to be able to pass through the low-pass filter. If this is the case the v.c.o. will continue to operate at its free-running frequency. If the difference frequency is somewhere in the vicinity of the filter's cut-off frequency some of its energy will be passed and some control applied to the v.c.o. This will move the v.c.o. frequency in the direction of the signal frequency and so reduce the difference frequency. This will allow more of the difference-frequency voltage to appear at the filter output and so on. This is a form of positive feedback and a cumulative action takes places that rapidly causes the p.l.l. to lock. The time taken for a p.l.l. to establish lock is known as the *pull-in time*, or as the *acquisition time*. Clearly, the pull-in time depends upon the initial frequency separation between the signal and v.c.o. frequencies, but it is also a function of both the loop gain and the bandwidth of the filter.

Analysis of a Phase-locked Loop

Figure 9.9 shows the block diagram of a p.l.l. with the conversion gain of each item, with voltages and frequencies marked.

Loop Gain

Once a p.l.l. has locked to an incoming signal the low-pass filter has an effective loss, or gain, of unity. The loop gain K_L is then

$$K_L = K_D K_0 A_v. \tag{9.12}$$

The units of K_L are V/rad $\times$ kHz/V $\times$ V/V or kHz/rad. The loop gain is numerically equal to the lock range.

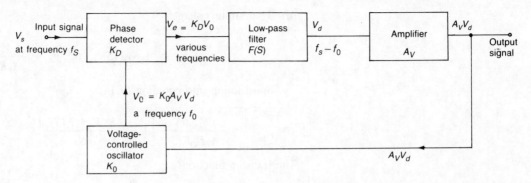

Fig. 9.9 Phase-locked loop.

If the frequency difference before lock is established is $f_s - f_c$ then a control voltage $A_V V_d = (f_s - f_c)/K_0$ is necessary to maintain lock. The phase detector must generate an error voltage $V_e = V_d = A_V V_d / A_V = (f_s - f_c)/K_0 A_V$ and this is derived from a static phase error $\theta_e = V_e / K_D$. Combining these two relationships gives

$$\theta_e = (f_s - f_c)/K_0 K_D A_V = (f_s - f_c)/K_L \text{ rad.} \tag{9.13}$$

Example 9.2

A p.l.l. has a phase detector with a conversion gain of $K_D = 0.5$ V/rad, an amplifier with a voltage gain of 6, and a v.c.o. of conversion gain $K_0 = 6$ kHz/V. If the free-running frequency of the v.c.o. is 90 kHz, calculate (a) the loop gain, (b) the v.c.o. frequency and the static phase error when the signal frequency is 100 kHz, and (c) the lock range.

Solution

(a) Loop gain $K_L = K_D K_0 A_V = 0.5 \times 6 \times 6 = 18$ kHz/rad. (*Ans.*)
(b) $f_c = 90$ kHz;
     $f_s - f_c = 10$ kHz; $A_V V_d = 10$ kHz/6 kHz/V $= 1.67$ V;
     $V_d = 1.67/6 = 0.28$ V; and
     $\theta_e = 0.28$ V/0.5 V/rad $= 0.56$ rad $= 32°$. (*Ans.*)
(c) Lock range Δf_L corresponds to $= \pm \pi/2$ rad.
Therefore, $\Delta f_L = K_L \pi = 56.55$ kHz. (*Ans.*)

Dynamic Behaviour

Once a p.l.l. has established lock with an incoming signal, non-linear capture transients no longer exist. The p.l.l. can then be regarded as a feedback system in which the static phase error $\theta_e = \theta_s - \theta_0$ is the variable. Referring to Fig. 9.9, the phase θ_s of the incoming signal is

$$\theta_s = \theta_e + \theta_0 = \theta_e + \frac{K_0 A_V V_d}{S} = \theta_e + \frac{K_D A_V K_0 A_V F(S) \theta_e}{S}$$

$$= \theta_e \left[1 + \frac{K_D A_V K_0 F(S)}{S} \right]. \tag{9.14}$$

(a) The output voltage

$$V_{out}(t) = A_V V_d = A_V K_D F(S)\theta_e \quad \text{or} \quad \theta_e = \frac{V_{out}(t)}{A_V K_D F(S)}.$$

Substituting into equation (9.14),

$$\theta_s = \frac{V_{out}(t)}{A_V K_D F(S)}\left[1 + \frac{K_D K_0 A_V F(S)}{S}\right].$$

The transfer function

$$T(S) = \frac{V_{out}(S)}{\theta_s(S)} = \frac{A_V K_D F(S)}{1 + \dfrac{K_D K_0 A_V F(S)}{S}}$$

$$= \frac{S A_V K_D F(S)}{S + K_L F(S)}. \tag{9.15}$$

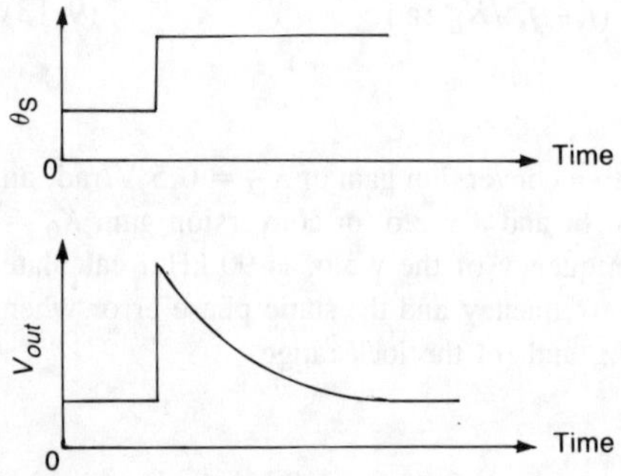

Suppose that the filter is not fitted so that $F(S) = 1$; then

$$T(S) = \frac{S A_V K_D}{S + K_L}. \tag{9.16}$$

The response of the p.l.l. to a change in the phase of the input signal can be determined using the Laplace transform. If, for example, the input phase should suddenly increase from one value to another — a *step function* — the change with time of the output voltage will be exponential as shown by Fig. 9.10.†

(b) If the frequency of the input signal changes

$$\frac{d\theta_s}{dt} = S\theta_s = \omega_s(s) \quad \text{and} \quad \frac{V_{out}(S)}{\theta_s} = \frac{V_{out}(S)}{\omega_s(S)/S} = \frac{SV_{out}(S)}{\omega_s(S)}.$$

From equation (9.16),

$$\frac{V_{out}(S)}{\omega_s(S)} = \frac{A_V K_D}{S + K_L}. \tag{9.17}$$

This transfer function is of the same form as that for the low-pass filter of Fig. 9.6(a) and so $V_{out}(t)$ changes with change in f_s in a similar manner to $V_{out}(t)/V_{in}(t)$ for the filter. For a step change in $f_s, \omega_s(s) = \Delta\omega_s/s$, and Laplace transform work shows

Fig. 9.10 Output voltage of p.l.l. for an abrupt change in input phase.

---

† For a step input, $\theta_s(s) = \theta_s/s$. Hence

$$V_{out}(s) = \frac{\theta_s}{S} \times \frac{S A_v K_D}{S + K_L} = \frac{\theta_s A_V K_D}{S + K_L}$$

and from tables $V_{out}(t) = \theta_s A_V K_D e^{-t/K_L}$.

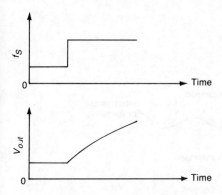

Fig. 9.11 Output voltage of p.l.l. for an abrupt change in input frequency.

Fig. 9.12 Bode diagram of uncompensated p.l.l.

that $V_{out}(t)$ rises exponentially with time, see Fig. 9.11. The rate of change of the output voltage may often be too low and the response time of the p.l.l. needs to be reduced. A reduction in the response time can be obtained by the use of *loop compensation*.

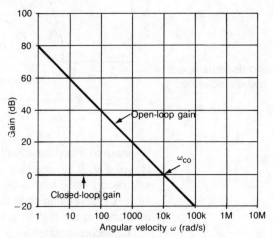

The Bode diagram of an uncompensated p.l.l. is given in Fig. 9.12; the loop gain K_L, quoted in dB, is plotted against frequency on a logarithmic scale. Suppose that the loop gain is 10 000 s⁻¹ at $\omega =$ 1 rad/s; then $20 \log_{10} 10\,000 = 80$ dB. The open-loop gain plot is a straight line starting at the point 80 dB, $\omega = 1$, and decreasing at 20 dB/decade. The plot passes through the 0 dB line at $\omega_{co} =$ 10 krad/s; note that this is numerically equal to the loop gain K_L. The closed-loop gain is 0 dB up to ω_{co} and then falls at 20 dB/decade. The error in the closed-loop plot is 3 dB at $\omega_{co} = K_L$ and less than this at all other frequencies. The p.l.l. can track signal frequencies that are either higher, or lower, than the free-running frequency of the v.c.o. and hence the bandwidth of the uncompensated p.l.l. is equal to $2K_L$.

Loop Compensation

The response time of a p.l.l. to a change in the input signal frequency can be reduced by the insertion into the feedback loop of a low-pass filter. The transfer functions of the loop with filter is given by equations (9.17) and (9.18).

$$T(S) = \frac{V_{out}(S)}{\omega_s(S)} = \frac{K_D A_V F(S)}{S + K_L F(S)}. \tag{9.18}$$

These are the transfer functions of second-order systems, and their

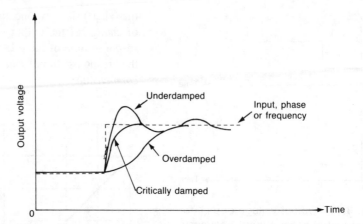

Fig. 9.13 Illustrating underdamped, critically damped, and over-damped p.l.l.s.

response to a step input will depend upon the *damping* of the loop. If, see Fig. 9.13, the loop is underdamped, overshoot and ringing will occur; if the loop is overdamped the transient response will be slow. Critical damping is the case where the output voltage is able to change at the highest possible rate *without* any overshoot.

The uncompensated loop has a loop phase shift of $180° - 90° = 90°$, and so it cannot oscillate. The low-pass filter will introduce extra phase shift, equal to $\tan^{-1}(\omega_{co}/\omega_3)$, where ω_3 is the cut-off frequency of the filter (Fig. 9.6(*a*)). The *phase margin* is the amount by which the loop phase shift at $\omega = K_L$ is less than $180°$.

Example 9.3

The p.l.l. of Example 9.2 has a low-pass filter with a cut-off frequency ω_3 = 5 krad/s. Draw the Bode diagram for the loop and estimate the phase margin.

Solution
The Bode plot is given in Fig. 9.14. The plot of the uncompensated loop falls at 20 dB/decade from 80 dB at $\omega = 1$ rad/s to $\omega_3 = 5$ krad/s. Above this frequency the low-pass filter causes the fall-off to increase to 40 dB/decade. The closed-loop 0 dB gain plot intersects with this line at ω_{co} = 6.5 krad/s. Hence the phase margin = $90° - \tan^{-1}(6.5/5) = 90° - 52°$ = $38°$.   (*Ans.*)

The phase margin will affect the frequency response of the p.l.l. Critical damping corresponds to a phase margin of $64°$; any phase margin smaller than this will produce a peak in the loop gain-frequency response. This point is illustrated by Fig. 9.15; a reduction in the phase margin increases the peak in the response and also increases overshoot and ringing in the response to a step input. Phase margins of less than $64°$ will produce an underdamped response. Should the phase margin be reduced to $0°$ the loop will oscillate at its *natural frequency* of $\omega_n = \sqrt{\omega_3 K_L}$.

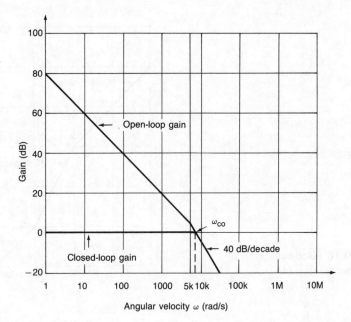

Fig. 9.14

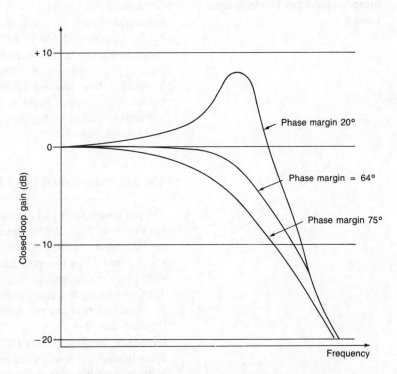

Fig. 9.15 Gain/frequency characteristics of p.l.l.s with different phase margins.

If the more complex filter of Fig. 9.6(b) is used the Bode diagram is as shown by Fig. 9.16. The bandwidth of the loop can now be set by choice of the time constant $C(R_1 + R_2) \simeq CR_1$ usually, and the damping can be set by choice of CR_2.

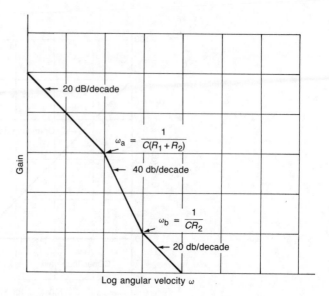

$$\omega_a = \frac{1}{C(R_1 + R_2)}$$

$$\omega_b = \frac{1}{CR_2}$$

Fig. 9.16 Bode diagram of compensated p.l.l.

Integrated-circuit Phase-locked Loops

Because of the complexity of the circuit requirements for a p.l.l. integrated circuits are nearly always employed. There are several ICs on the market, and one of these, the 565, has become an industry standard that can be employed for a variety of purposes. Other p.l.l.s such as, for example, the Signetics NE560/1/2/4 or the Plessey SL650/1/2, have features that make them especially suitable for particular applications. Since it is an industry standard device and, although originated by Signetics, is now second-sourced, the 565 p.l.l. will be considered.

The 565 Phase-locked Loop Integrated Circuit

The pin connections and the internal block diagram of the 565 p.l.l. are shown by Figs. 9.17(a) and (b), respectively. It can be seen that the IC includes the v.c.o., the phase detector, the amplifier and the resistor part of the low-pass filter. The low-pass filter must be completed by the connection of a capacitor C_2 between pins 7 and 10. The free-running frequency of the v.c.o. is set by the value of a resistor R_1 connected between pin 8 and $+V_{CC}$, and a capacitor connected between pin 9 and $-V_{CC}$. The 565 p.l.l. can operate over a frequency band of 0.001 Hz to 500 kHz and generates both a demodulated and a reference output voltage. The reference voltage is made available so that, if required, the p.l.l. demodulated voltage can be applied to a voltage comparator. The v.c.o. output is brought out to pin 4 to provide a TTL-compatible square-wave voltage waveform. In most applications of the IC pin 4 must be connected to pin 5 to provide the phase detector with a v.c.o. input voltage.

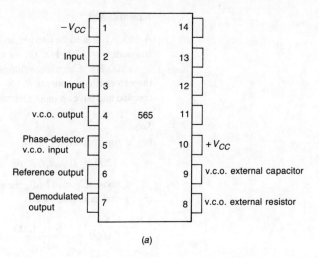

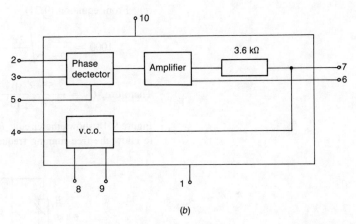

Fig. 9.17 565 p.l.l. (a) pin connections (b) internal block diagram.

The two input pins 2 and 3 must be given identical d.c. bias voltages, somewhere in the range 0 V to -4 V. The IC has, typically, $K_D = 0.45$ V/rad, $K_O = 6.6$ kHz/V and $A_V = 1.4$. The manufacturer's data sheet gives the following design fomulae.

(a) Free-running frequency of v.c.o. $f_c = 0.3/R_1C_1$ (9.19)

    C_1 can be any value but R_1 must have a value within the range 2000 Ω to 20 kΩ, with a value near to 4000 Ω being preferred.

(b) Lock range $f_L = \pm 8f_cV_{CC}$ Hz. (9.20)

(c) Capture range $\pm \dfrac{1}{2\pi}\sqrt{\dfrac{2\pi f_L}{\tau}}$ (9.21)

    where $\tau = 3.6 \times 10^3 \, C_2$ s.

(d) Loop gain $= 50 \, f_c/V_{CC}$. (9.22)

    The loop gain can be reduced by connecting a resistor R_3 between pins 6 and 7 to reduce the gain A_V of the internal amplifier.

Example 9.4

A 565 p.l.l. is to be used in an application which requires (*a*) free-running frequency = 1200 Hz, (*b*) lock range = ±800 Hz, and (*c*) capture range = ±500 Hz. Calculate suitable values for the supply voltage ±V_{CC}, and the external components R_1, R_2 and C_1. The input of the p.l.l. is to be a.c. coupled and have an input resistance of about 600 Ω. Draw the circuit diagram.

Solution
(*a*) From equation (9.19),

$$1200 = 0.3/R_1C_1 \quad \text{or} \quad R_1C_1 = 0.25 \times 10^{-3} \text{ s.}$$

Choose C_1 = 47 nF, then R_1 = 5319 Ω.
(*b*) From equation (9.20)

$$1600 = \frac{16 \times 1200}{V_{CC}} \quad \text{or} \quad V_{CC} = \pm 12 \text{ V.}$$

(*c*) From equation (9.21)

$$1000 = \frac{1}{2\pi} \sqrt{\frac{2\pi \times 1600}{\tau}} \quad \text{or} \quad \tau = 2.55 \times 10^{-4} \text{ s.}$$

Therefore, $C_2 = \dfrac{2.55 \times 10^{-4}}{3.6 \times 10^3} = 70.8$ nF.

Figure 9.18 shows the circuit. R_1 has been made a variable 6.8 kΩ resistor to allow the free-running frequency of the v.c.o. to be adjusted.

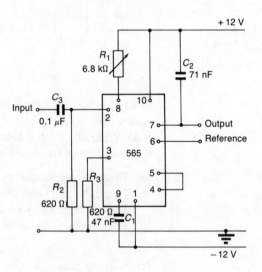

Fig. 9.18

Phase-locked Loop Applications

A p.l.l. can be employed for a variety of applications in which there is a requirement for a signal to be demodulated, or to be tracked, for a frequency to be multiplied, or for a stable source of many frequencies.

Demodulation of Frequency-modulated Signals

The use of a p.l.l. as an f.m. demodulator requires only the basic circuit such as, for example, that given in Fig. 9.18. The incoming signal will need to be amplified and band-limited, and often shifted to a lower frequency, before it is applied to the p.l.l. The free-running frequency should be set to the centre frequency of the signal to be demodulated.

For the demodulation of an f.s.k. signal a p.l.l. with both demodulated and reference outputs will be necessary; these outputs will be connected to the two inputs of a voltage comparator. The comparator should then produce a TTL or CMOS compatible output waveform.

Frequency Multiplication

A p.l.l. can be employed as a frequency multiplier by locking it to a harmonic of the input signal frequency. A better method is illustrated by Fig. 9.19. The v.c.o. is set to have its free-running frequency at about the required harmonic of the input signal frequency and so the loop locks at that frequency. The output of the v.c.o. is not connected to the phase detector but, instead, it is connected to a digital divide-by-N circuit. The division ratio N is chosen to be equal to the order of the signal-frequency harmonic at which the v.c.o. is operating.

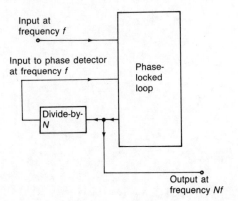

Fig. 9.19 Frequency multiplier.

The output of the divider is applied to the phase detector and causes the system to lock to the required harmonic. If, for example, the input signal is from a very stable crystal oscillator operating at 50 kHz and $N = 75$ then the output signal frequency will be 75×50 kHz $= 3.75$ MHz. The output frequency will be as stable as the reference crystal oscillator frequency.

Stable Reference Frequency

A p.l.l. can be locked to a standard-frequency radio signal, such as the Droitwich 200 kHz transmission. The signal must be picked up by an aerial, probably amplified, and applied to the p.l.l. The output from the v.c.o. can then be multiplied and/or divided to obtain frequencies with the same degree of high accuracy as the standard r.f. transmission.

10 Active Filters

An active filter is a circuit that using one, or more, op-amps, capacitors and resistors but *no* inductance, provides a frequency-selective transfer function. An active filter will pass, with little attenuation, a certain band of frequencies whilst introducing considerable attenuation at all other frequencies outside of that band. Four basic types of filter are possible; these are the low-pass, the high-pass, the band-pass and the band-stop filters. Figure 10.1 shows the idealized attenuation/frequency characteristics of each. Each filter has a defined *passband* in which the attenuation is zero, and a defined *stopband* in which the attenuation is large. The change from passband to stopband cannot be instantaneous and the frequency interval in which the changeover takes place is known as the *transition band*. The narrower the transition band the more *selective* is the filter. The rate at which the attenuation increases with change in frequency is known as the *roll-off* of the filter.

There are two main ways of designing a filter known, respectively, as the *Butterworth* and the *Tchebysheff* filters. The differences between these designs are illustrated, for the low-pass filter, by the attenuation/frequency characteristics given in Fig. 10.2. The Butterworth filter, Fig 10.2(*a*), has a *maximally flat* response in the passband; this means that the response is as flat as possible over the low-frequency end of the passband and that there is no ripple in the passband. This is achieved at the expense of an indifferent selectivity. The characteristic is *monotonic*, i.e. an increase in frequency always produces an increase in the attenuation. The passband is specified by the *cut-off frequency* f_c; this is the frequency at which the filter has an attenuation of 3 dB. The frequency f_s is the frequency at which the minimum attenuation A dB, which specifies the stopband, is reached. The Tchebysheff filter, Fig 10.2(*b*), has its attenuation increase more rapidly outside of the passband than does the Butterworth filter, but this is at the expense of *ripple* in the passband. The greater the ripple that is allowed the more selective will be the filter. The end of the passband of a Tchebysheff filter is usually taken as being the maximum (for a low-pass filter) frequency f_r at which the passband attenuation is at its maximum value; i.e. when the highest-frequency peak ripple occurs. The frequency f_r is known as

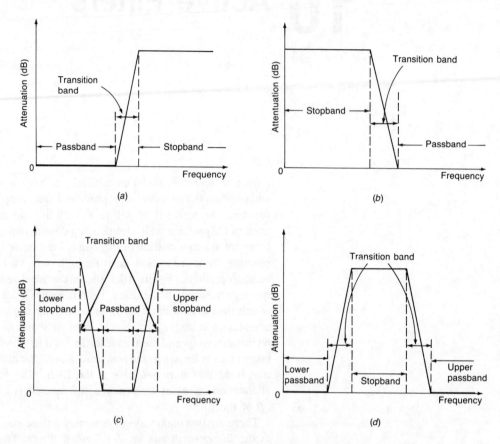

Fig. 10.1 Ideal attenuation/frequency characteristics for
(a) low-pass, (b) high-pass, (c) band-pass, and (d) band-stop filters.

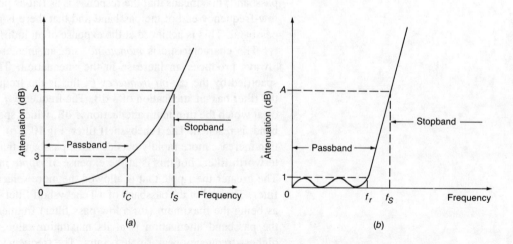

Fig. 10.2 (a) Butterworth and (b) Tchebysheff low-pass filter
attenuation/frequency characteristics.

the *ripple bandwidth*. For example, if a Tchebysheff filter has 1 dB ripple and $f_r = 10$ kHz then its response is ± 1 dB from 0 Hz to 10 kHz with a rapidly increasing attenuation at frequencies greater than 10 kHz. The minimum attenuation A specified for the stopband is attained at frequency f_s.

First-order Filters

A first-order filter is one whose transfer function contains no terms in S^2. Figure 10.3(a) shows an RC first-order low-pass filter; its transfer function was shown on page 215 to be given by $F(S) = 1/(1+SCR)$. The designed-for performance of such a circuit will be modified by the resistance of the load to which it is connected and so it is often followed by a voltage follower, see Fig. 10.3(b). An alternative circuit is given in Fig. 10.3(c); this circuit has a transfer function $F(S) = V_{out}(S)/V_{in}(S)$ of

$$F(S) = \frac{-R_2/(1+SC\,R_2)}{R_1} = \frac{-R_2/R_1}{1+SCR_2}.$$

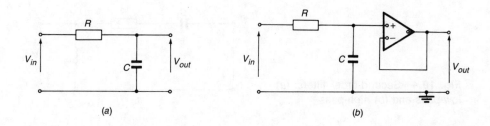

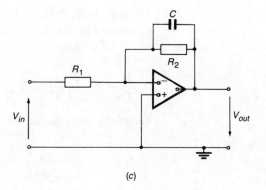

Fig. 10.3 First-order filters: (a) basic circuit; (b) buffered circuit; and (c) op-amp circuit.

The attenuation of a first-order filter increases at the rate of 6 dB/octave or 20 dB/decade. Very often this selectivity is inadequate and greater selectivity can be obtained by cascading a number of circuits. A better alternative, however, is to employ a *second-order filter*.

Second-order Filters

Figure 10.4(a) shows a passive second-order filter whose selectivity and cut-off frequency are determined by the values of its inductance and capacitance. The transfer function $F(S)$ of the filter is

$$F(S) = \frac{V_{out}(S)}{V_{in}(S)} = \frac{1/SC}{SL + R + \dfrac{1}{SC}} = \frac{1}{S^2 LC + SCR + 1}$$

$$F(S) = \frac{1/LC}{S^2 + \dfrac{SR}{L} + \dfrac{1}{LC}}. \tag{10.1}$$

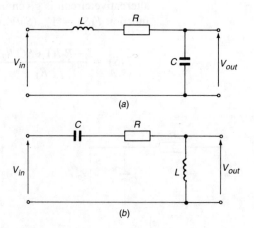

(a)

(b)

Fig. 10.4 Second-order filters: (a) low-pass and (b) high-pass.

The final term, $1/LC$ in the denominator of equation (10.1) is the square of (2π times the resonant frequency f_0 of the circuit), i.e. $\omega_0^2 = 1/LC$. Also,

$$\frac{R}{L} = R\sqrt{\frac{C}{L^2 C}} = R\sqrt{\frac{C}{L}} \cdot \frac{1}{\sqrt{LC}} = \omega_0/Q.$$

Therefore, the equation for $F(S)$ can be rewritten as

$$F(S) = \frac{\omega_0^2}{S^2 + \dfrac{S\omega_0}{Q} + \omega_0^2} \tag{10.2}$$

$$\text{or } F(S) = \frac{1}{\dfrac{S^2}{\omega_0^2} + \dfrac{S}{\omega_0 Q} + 1}. \tag{10.3}$$

For the second-order high-pass filter of Fig. 10.4(b),

$$F(S) = \frac{V_{out}(S)}{V_{in}(S)} = \frac{SL}{SL+R+\dfrac{1}{SC}} = \frac{S}{S+\dfrac{R}{L}+\dfrac{1}{SCL}}$$

$$= \frac{S^2}{S^2+\dfrac{SR}{L}+\dfrac{1}{LC}}$$

or $F(S) = \dfrac{S^2}{S^2+\dfrac{S\omega_0}{Q}+\omega_0^2}$. $\qquad$ (10.4)

It should be noted that the two transfer functions have the same denominator. If the transfer function for a bandpass filter, or for a bandstop filter, is determined the results are

(*a*) bandpass filter

$$F(S) = \frac{S}{S^2+\dfrac{S\omega_0}{Q}+\omega_0^2}, \qquad (10.5)$$

(*b*) bandstop filter

$$F(S) = \frac{(S^2+\omega_0^2)}{S^2+\dfrac{S\omega_0}{Q}+\omega_0^2}. \qquad (10.6)$$

It is often convenient to employ the *damping factor* $\delta = 1/2Q$ in the common denominator of the four filter transfer functions.

The Butterworth Filter

The transfer function of a Butterworth low-pass filter is written as $F(S) = H_0/B_n(S)$, where H_0 is the d.c. gain, and $B_n(S)$ is the polynomial for the nth-order filter. For a sinusoidal signal $B_n(S) = B_n(j\omega)$ and its magnitude is given by

$$|B_n(\omega)| = \sqrt{\left[1+\left(\frac{\omega}{\omega_c}\right)^{2n}\right]} \qquad (10.7)$$

where n is the order of the filter. Hence the magnitude of the transfer function is

$$|F(j\omega)| = \frac{1/\omega_c^2}{\sqrt{\left[1+\left(\dfrac{\omega}{\omega_c}\right)^{2n}\right]}} \qquad (10.8)$$

If $\omega = \omega_c$, then

$$\left| \frac{F(j\omega)}{1/\omega_c^2} \right| = \frac{1}{\sqrt{1+(\omega/\omega_c)^{2n}}} = \frac{1}{\sqrt{2}}$$

whatever the value of n. This means that the output voltage of a Butterworth low-pass filter will be 3 dB down on its maximum value at $\omega = \omega_c$ no matter what the order of the filter may be. If the frequency is doubled, so that $\omega = 2\omega_c$ then the response of the filter will depend upon its order n, see Table 10.1.

Table 10.1

n	$\left(\dfrac{\omega}{\omega_c}\right)^{2n}$	$1+\left(\dfrac{\omega}{\omega_c}\right)^{2n}$	$\sqrt{\left[1+\left(\dfrac{\omega}{\omega_c}\right)^{2n}\right]}$	$\dfrac{V_{out}}{V_{in}}$ (dB)
1	$2^2 = 4$	5	2.336	-7
2	$2^4 = 16$	17	4.123	-12
3	$2^6 = 64$	65	8.062	-18
4	$2^8 = 256$	257	16.031	-24

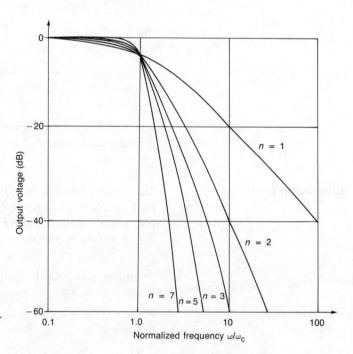

Fig. 10.5 Butterworth low-pass filter characteristic.

The data given in Table 10.1, plus higher frequency ratios, is shown plotted in Fig. 10.5. Clearly, the higher the order of the filter the flatter is the passband response, and the faster is the roll-off, i.e. the more nearly is the ideal response approached. There is no passband ripple.

Example 10.1

A Butterworth low-pass filter is required to have a loss of 50 dB at a frequency three times the cut-off frequency. Find the necessary order of the filter.

Solution

$\omega/\omega_c = 3$. Therefore, $(\log_{10}^{-1} 2.5)^2 = 10^5 = 1+3^{2n}$

$2n \log_{10}3 = 0.9542$ $n = \log_{10}10^5 = 5$

$$n = \frac{5}{0.9542} = 5.24.$$

So a sixth-order filter is required. (*Ans.*)

Consider a first-order filter, Fig. 10.3(*a*), which has $F(S) = 1/(1+SCR)$. If the cut-off frequency $\omega_c = 1/CR$ is normalized to 1 rad/s, $F(S) = 1/(1+S)$. Therefore,

$$B_n(S) = 1+S.$$

For second-, and higher-order, filters the factors of $B_n(S)$ can be determined using equation (10.9), i.e.

$$n\text{th-order factors} = 2 \sin\left[\frac{(2k-1)\pi}{2n}\right]. \tag{10.9}$$

All odd-order polynomials have a factor $(S+1)$. Consider $n = 4$;

$$2\sin\left[\frac{(2k-1)\pi}{8}\right] : \text{if } k = 1 \text{ gives } 2 \sin 22.5° = 0.765,$$
$$\text{if } k = 2 \text{ gives } 2 \sin 67.5° = 1.848$$
$$\text{if } k = 3 \text{ gives } 2 \sin 112.5° = 1.848 \text{ } again \text{ so}$$
$$\text{there are no further factors.}$$

Carrying on in this way allows a table of normalized Butterworth polynomials to be obtained; this is given by Table 10.2 for $n = 1$ to $n = 6$.

In practice, because of component tolerances and changes in component value with time the Butterworth filter will always show some

Table 10.2

n	Factors of $B_n(S)$
1	$S + 1$
2	$S^2 + 1.414 + 1$
3	$(S + 1)(S^2 + 1.414 + 1)$
4	$(S^2 + 0.765S + 1)(S^2 + 1.848S + 1)$
5	$(S + 1)(S^2 + 0.765S + 1)(S^2 + 1.848S + 1)$
6	$(S^2 + 0.518S + 1)(S^2 + 1.414S + 1)(S^2 + 1.932S + 1)$

variation in its passband response, particularly near to the cut-off frequency.

The Tchebysheff Filter

The maximally flat approximation to a low-pass filter characteristic given by a Butterworth filter is good at frequencies in the region of zero hertz but it is much poorer nearer to the edge of the passband. If some ripple can be tolerated within the passband a filter with a more selective characteristic can be designed. This network is known as a Tchebysheff filter and it has a transfer function of the form given in equation (10.10).

$$\left|\frac{H(j\omega)}{H_0}\right| = \frac{1}{\sqrt{\left[1+\epsilon^2 C_n^2\left(\dfrac{\omega}{\omega_r}\right)\right]}} \qquad (10.10)$$

where $C_n(\omega/\omega_r)$ is a Tchebysheff polynomial,

$$C_n(\omega/\omega_r) = \cos\left(n\cos^{-1}\frac{\omega}{\omega_r}\right) \quad \text{for} \ \ 0 \le \frac{\omega}{\omega_c} \le 1 \quad (10.11(a))$$

$$= \cosh\left(n\cosh^{-1}\frac{\omega}{\omega_r}\right) \quad \text{for} \ \ \frac{\omega}{\omega_c} \ge 1 \quad (10.11(b))$$

(where $\cosh x = (e^x+e^{-x})/2$)

$$\simeq 2^{n-1}\left(\frac{\omega}{\omega_r}\right)^n \quad \text{if} \ \frac{\omega}{\omega_c} \gg 1 \qquad (10.12)$$

and $\epsilon = \sqrt{(10^{\gamma/10}-1)}$,

where γ = passband ripple in dB.
Thus for 1 dB ripple $\epsilon = \sqrt{(10^{0.1}-1)} = 0.509$ and for 0.5 dB ripple $\epsilon = \sqrt{(10^{0.05}-1)} = 0.3493$.

For a 3 dB fall in the output voltage $\left|\dfrac{H(j\omega_3)}{H_0}\right|^2 = 1/2$, so

$$2 = 1 + \epsilon^2 C_n^2\left(\frac{\omega_3}{\omega_r}\right), \ 1 = \epsilon\cosh\left(n\cosh^{-1}\frac{\omega_3}{\omega_r}\right)$$

$$\cosh^{-1}(1/\epsilon) = n\cosh^{-1}\frac{\omega_3}{\omega_r}, \ \text{or} \ f_3 = f_r\cosh\left[\frac{1}{n}\cosh^{-1}\left(\frac{1}{\epsilon}\right)\right].$$

$$\qquad (10.13)$$

Example 10.2

A Tchebysheff fourth-order low-pass filter has a ripple bandwidth of 3 kHz and 0.5 dB ripple in the passband. Calculate its 3 dB bandwidth.

Solution

$\epsilon = 0.3493$; hence from equation (10.13)

$f_3 = 3000 \cosh (0.25 \cosh^{-1} 2.862) = 3000 \cosh(0.25 \times 1.713)$

$= 3279$ Hz.   (*Ans.*)

Tchebysheff Polynomials

The Tchebysheff polynomials of order n are given by equations (10.11(a)) and (10.11(b)).

For $n = 1$, $C_n\left(\dfrac{\omega}{\omega_r}\right) = \cos \left(\cos^{-1} \dfrac{\omega}{\omega_r}\right) = \dfrac{\omega}{\omega_r}$.

For $n = 2$, $C_n\left(\dfrac{\omega}{\omega_r}\right) = \cos\left(2\cos^{-1} \dfrac{\omega}{\omega_r}\right) = 2\cos^2\left(\cos^{-1}\dfrac{\omega}{\omega_r}\right) - 1$

$= 2\left(\dfrac{\omega}{\omega_r}\right)^2 - 1.$

For $n = 3$, $C_n\left(\dfrac{\omega}{\omega_r}\right) = \cos\left(3\cos^{-1}\dfrac{\omega}{\omega_r}\right)$

$= 4\cos^3\left(\cos^{-1}\dfrac{\omega}{\omega_r}\right) - 3\cos\left(\cos^{-1}\dfrac{\omega}{\omega_r}\right)$

$= 4\left(\dfrac{\omega}{\omega_r}\right)^3 - 3\,\dfrac{\omega}{\omega_r}.$

For higher orders of n, use

$$C_n\left(\dfrac{\omega}{\omega_r}\right) = 2\dfrac{\omega}{\omega_r} C_{n-1}\left(\dfrac{\omega}{\omega_r}\right) - C_{n-2}\left(\dfrac{\omega}{\omega_r}\right) \text{ etc.} \qquad (10.14)$$

Example 10.3

Use equation (10.14) to obtain $C_4\left(\dfrac{\omega}{\omega_r}\right)$.

Solution

$$C_4\left(\dfrac{\omega}{\omega_r}\right) = 2\left(\dfrac{\omega}{\omega_r}\right) C_3\left(\dfrac{\omega}{\omega_r}\right) - C_2\left(\dfrac{\omega}{\omega_r}\right)$$

$$= 2\left(\dfrac{\omega}{\omega_r}\right) \left[4\left(\dfrac{\omega}{\omega_r}\right)^3 - 3\left(\dfrac{\omega}{\omega_r}\right)\right] - 2\left(\dfrac{\omega}{\omega_r}\right)^2 + 1$$

or $C_4\left(\dfrac{\omega}{\omega_r}\right) = 8\left(\dfrac{\omega}{\omega_r}\right)^4 - 8\left(\dfrac{\omega}{\omega_r}\right)^2 + 1.$ (*Ans.*)

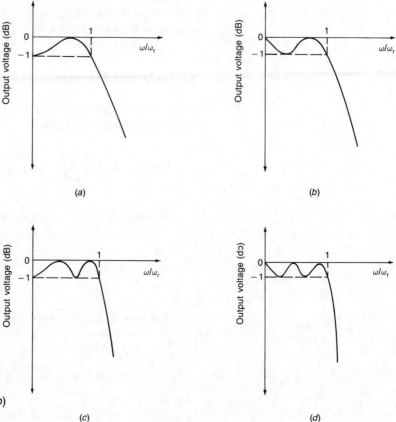

Fig. 10.6 Tchebysheff low-pass filter characteristics: (a) n = 2, (b) n = 3, (c) n = 4, and (d) n = 5.

If $C_n(\omega/\omega_r)$ is plotted over the passband, i.e. $\omega < \omega_r$, the passband ripple for various values of n can be seen, see Fig. 10.6. It should be noted that the number of ripple peaks in each passband is equal to the order n of the filter.

The attenuation of the filter outside of the passband can be obtained from the denominator of equation (10.10), i.e. $10 \log_{10} [1+\epsilon^2 C_n^2 (\omega/\omega_r)]$, where $\omega > \omega_r$. Suppose, for example, that n = 3, γ = 1 dB and $\omega = 2\omega_r$. Then

$$C_n\left(\frac{\omega}{\omega_r}\right) = \cosh(3\cosh^{-1}2) = \cosh 3.951 = 26$$

and the attenuation is $10 \log_{10} (1+0.5089^2 \times 26^2) = 22.46$ dB.

Example 10.4

A Tchebysheff low-pass filter is required to have 1 dB passband ripple and 50 dB attenuation at a frequency twice the ripple bandwidth. Determine the required order of the filter.

Solution

Required voltage ratio at $f = 2f_r = \log_{10}^{-1} (50/20)$. Therefore, $(\log_{10}^{-1} 2.5)^2$
$= 0.5089^2 C_n^2(2)$.

$$C_n^2(2) = \sqrt{\frac{100\,000}{0.5089^2}} = 621.4 = \cosh\,(n\cosh^{-1}2).$$

Try $n = 4$, $\cosh 5.268 = 194$, which is too small.
Try $n = 5$, $\cosh 6.585 = 724$. Hence the order of the required filter is 5.

(*Ans.*)

The design of an active Tchebysheff filter makes use of tables of the factors of the Tchebysheff polynomials. The formulae for these are much more complex than those for the Butterworth filter and so are not given here. Table 10.3 give the factors for passband ripples of 0.5 dB and 1 dB; in specialist filter books figures are given for ripples of many other magnitudes.

Table 10.3 (*a*) 0.5 dB ripple

n	*Factors*
1	$S + 2.863$
2	$S^2 + 1.425\,S + 1.516$
3	$(S + 0.626)\,(S^2 + 0.626S + 1.142)$
4	$(S^2 + 0.351S + 1.064)\,(S^2 + 0.845S + 0.356)$
5	$(S + 0.362)\,(S^2 + 0.224S + 1.036)\,(S^2 + 0.586S + 0.477)$
6	$(S^2 + 0.155S + 1.024)\,(S^2 + 0.414S + 0.548)$ $(S^2 + 0.580S + 0.157)$

(*b*) 1 dB ripple

n	*Factors*
1	$S + 1.965$
2	$S^2 + 1.098S + 1.103$
3	$(S + 0.494)\,(S^2 + 0.494S + 0.994)$
4	$(S^2 + 0.279S + 0.987)\,(S^2 + 0.674S + 0.279)$
5	$(S + 0.289)\,(S^2 + 0.179S + 0.989)\,(S^2 + 0.469S + 0.429)$
6	$(S^2 + 0.124S + 0.991)\,(S^2 + 0.340S + 0.558)$ $(S^2 + 0.464S + 0.125)$

Active Filters

A large number of designs for active filters have been proposed in the literature but only three; Sallen and Key types, state variable filters, and networks using gyrators, have been much used. The majority of active filters are probably of the Sallen and Key type and hence only these will be discussed in any detail.

Sallen and Key Filters

The general form of a Sallen and Key filter is shown in Fig. 10.7. The resistors R_{f1} and R_{f2} set the voltage gain $A_{v(f)}$ of the op-amp to the desired value.

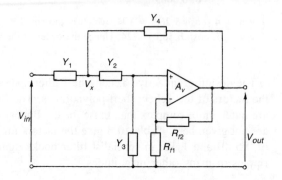

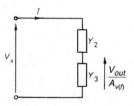

Fig. 10.8

Fig. 10.7 Active filter of the Sallen and Key type.

Summing the currents at the node A:

$$(V_{in} - V_x)Y_1 + (V_{out} - V_x)Y_4 - \frac{V_{out}Y_3}{A_{v(f)}} = 0$$

$$V_{in}Y_1 = (Y_1 + Y_4)V_x - V_{out}Y_4 + \frac{V_{out}Y_3}{A_{v(f)}}.$$

Referring to Fig. 10.8, $I = V_{out}Y_3/A_{v(f)}$ and so

$$V_x = \frac{V_{out}Y_3}{A_{v(f)}}\left(\frac{Y_2 + Y_3}{Y_2 Y_3}\right) = \frac{V_{out}}{A_{v(f)}}\left(\frac{Y_2 + Y_3}{Y_2}\right).$$

Hence

$$V_{in}Y_1 = \left[(Y_1 + Y_4)\left(\frac{Y_2 + Y_3}{Y_2}\right) + Y_3\right]\frac{V_{out}}{A_{v(f)}} - V_{out}Y_4$$

or $F(S) = \dfrac{V_{out}}{V_{in}} = \dfrac{A_{v(f)}\,Y_1 Y_2}{(Y_1 + Y_4)(Y_2 + Y_3) + Y_2 Y_3 - A_{v(f)}Y_2 Y_4}$

$$= \frac{A_{v(f)}\,Y_1 Y_2}{Y_2[Y_1 + Y_4(1 - A_{v(f)})] + Y_3(Y_1 + Y_2 + Y_4)}. \tag{10.15}$$

For the network to act as a low-pass, or a high-pass, or a band-pass filter the admittances Y_1, Y_2, etc. must be provided by the appropriate components (R or C) so that the transfer function has the form of equation (10.1), etc.

Suppose a low-pass filter is wanted. Then, since the numerator of equation (10.1) does not contain a term in S, both Y_1 and Y_2 must be provided by resistors. Equation (10.15) then reduces to

$$F(S) = \frac{A_{v(f)} \, G_1 G_2}{G_2[G_1 + Y_4(1 - A_{v(f)})] + Y_3(G_1 + G_2 + Y_4)} \qquad (10.16)$$

and from this it can be seen that for terms in both S and S^2 to appear in the denominator both Y_3 and Y_4 must be provided by capacitors. The required circuit is shown by Fig. 10.9.

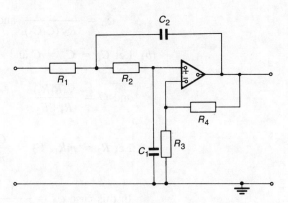

Fig. 10.9 Low-pass active filter.

The transfer function of Fig. 10.9 can be derived directly or by substituting into equation (10.16), thus

$$F(S) = \frac{A_{v(f)} \, G_1 G_2}{G_1 G_2 + SC_2 G_2(1 - A_{v(f)}) + SC_1(G_1 + G_2 + SC_2)}$$

$$= \frac{A_{v(f)}}{1 + SC_2 R_1(1 - A_{v(f)}) + SC_1[R_1 + R_2 + SC_2 R_1 R_2]}$$

$$\text{or } F(S) = \frac{A_{v(f)}}{S^2 C_1 C_2 R_1 R_2 + S[C_2 R_1(1 - A_{v(f)}) + C_1(R_1 + R_2)] + 1} \qquad (10.17)$$

Comparison of equation (10.17) with equation (10.3) gives

$$\omega_0 = \frac{1}{\sqrt{(R_1 R_2 C_1 C_2)}} \quad \text{and} \quad Q = \frac{\sqrt{(R_1 R_2 C_1 C_2)}}{R_1 C_2(1 - A_{v(f)}) + C_1(R_1 + R_2)}.$$

Often the design of a particular filter commences with making $R_1 = R_2$ and $C_1 = C_2$. Then

$$F(S) = \frac{A_{v(f)}}{S^2 R^2 C^2 + SRC \,(3 - A_{v(f)}) + 1}. \qquad (10.18)$$

Also, $\omega_0 = \dfrac{1}{RC}$ and $Q = \dfrac{1}{3 - A_{v(f)}}$. $\qquad (10.19)$

The gain $A_{v(f)}$ of the op-amp must be less than 3, otherwise the circuit will be unstable. The Q-factor can be adjusted by varying one resistor (R_1 or R_2) without affecting ω_0.

Alternative approaches to a design are as follows.

(a) Let $R_1 = R_2 = R$ and let $A_{v(f)} = 1$. It will then be found that

$$\omega_0 = \frac{1}{R\sqrt{(C_1 C_2)}} \quad \text{and} \quad Q = \frac{1}{2}\sqrt{\frac{C_2}{C_1}}.$$

(b) Let $C_1 = C_2 = C$ and $A_{v(f)} = 1$. Then $\omega_0 = \dfrac{1}{C\sqrt{(R_1 R_2)}}$

and $Q = \dfrac{\sqrt{(R_1 R_2)}}{R_1 + R_2}$.

(c) Let $R_2 = mR_1$, $C_2 = \dfrac{C_1}{m}$ and $A_{v(f)} = 1$.

In this case $\omega_0 = \dfrac{1}{R_1 C_1}$ and $Q = \dfrac{m}{1+m}$.

It can be seen that in each of these cases ω_0 and Q are interdependent and so cannot be separately adjusted. For this reason the first design is the most often employed and only it is further considered in this chapter.

For a Butterworth filter all sections have the same values of resistance and capacitance since ω_0 is the same throughout. Each section will, however, have a different gain $A_{v(f)}$.

The circuit of Fig. 10.9 provides a second-order filter. Higher-order filters are obtained by cascading first- and second-order filter sections; thus a third-order filter is obtained by cascading a first-order filter section with a second-order section, a fourth-order filter by cascading two second-order filters, and so on.

Example 10.5

Design a second-order low-pass Butterworth filter to have a cut-off frequency of 3 kHz.

Solution
From Table (10.2) the factors of $B_n(S)$ are $S^2 + 1.414S + 1$. Hence, comparing with equation (10.19),

$$\frac{1}{1.414} = \frac{1}{3 - A_{v(f)}} \quad \text{or} \quad A_{v(f)} = 3 - 1.414 = 1.586.$$

If R_3 is chosen to be 10 kΩ R_4 will be 5.86 kΩ. Also, $f_0 = 3000 = 1/2\pi RC$. Choosing the convenient value of 0.022 μF (22 nF) for $C_1 = C_2$ gives $R_1 = R_2 = 2410\ \Omega$. Figure 10.10 shows the circuit of the filter.

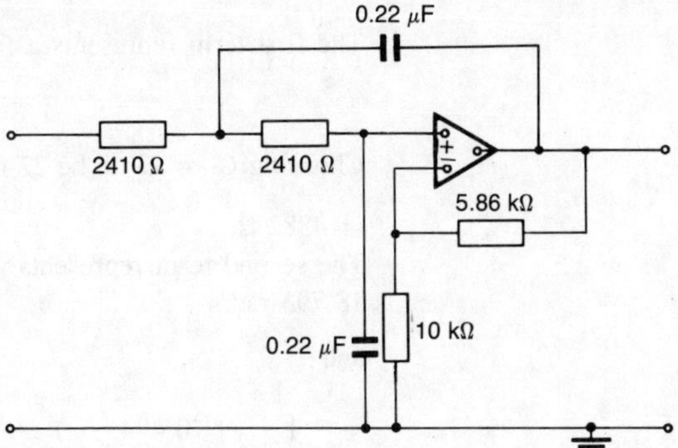

Fig. 10.10

Example 10.6

Design a second-order low-pass Tchebysheff filter to have a ripple bandwidth of 3 kHz and 0.5 dB ripple in the passband.

Solution

From Table 10.3 the factors are $S^2 + 1.425S + 1.516$. Rewriting in the same form as equation (10.18) and denormalizing gives

$$\frac{S^2}{1.516\omega_r^2} + \frac{1.425S}{1.516\omega_r} + 1.$$

Comparing equations, with $\omega_r = 2\pi \times$ required ripple frequency,

$$\omega_0 = \omega_r\sqrt{1.516} = 1.23\omega_r = 1.23 \times 2\pi \times 3000 = 23\ 185 \text{ rad/s},$$

and $\dfrac{1}{\omega_0 Q} = \dfrac{1.425}{1.516\omega_r}, \dfrac{1}{Q} = \dfrac{1.425\omega_0}{1.516\omega_r} = \dfrac{1.425\omega_r\sqrt{1.516}}{1.516\omega_r} = \dfrac{1.425}{\sqrt{1.516}}$

or $Q = 0.864$.

$$A_{v(f)} = 3 - \frac{1}{0.864} = 1.843. \text{ If } R_3 = 10 \text{ k}\Omega \text{ then } 1.843 = 1 + \frac{R_4}{10}$$

or $R_4 = 8.43$ kΩ. Choosing $C_1 = C_2$ as 22 nF, $R_1 = R_2 = \dfrac{10^6}{23\ 185 \times 0.022}$

$= 1960\ \Omega$.

Example 10.7

Design a third-order Tchebysheff low-pass filter with 1 dB passband ripple and 3 kHz ripple bandwidth.

Solution

From Table 10.3 the factors are $(S+0.494)(S^2+0.494S+0.994)$. Rewriting and denormalizing gives

$$\frac{S}{0.494\omega_r} + 1 \text{ and } \frac{S^2}{0.994\omega_r} + \frac{0.494S}{0.994\omega_r} + 1.$$

The first term represents a first-order filter with $RC = \dfrac{1}{0.494\omega_r}$.

Choosing $C_1 = C_2$ to be 22 nF gives $R_1 = R_2 = \dfrac{10^9}{0.494\times 2\pi \times 3000 \times 22}$
$= 4882 \ \Omega$.

The second term represents a second-order filter with $\omega_0 = \omega_r\sqrt{0.994} = 18\ 793$ rad/s

and

$$\frac{1}{\omega_0 Q} = \frac{0.494}{0.994\omega_r}, \frac{1}{Q} = \frac{\sqrt{0.994}\times 0.494\omega_r}{0.994\omega_r} = \frac{0.494}{\sqrt{0.994}} = 0.496$$

and

$$Q = 2.02.$$

Therefore, $A_{v(f)} = 3 - 0.496 = 2.504$. If $R_3 = 10$ kΩ then $R_4 - 15.04$ kΩ. Choosing $C_1 = C_2 = 22$ nF,

$$R_1 = R_2 = \frac{10^9}{18\ 793 \times 22} = 2419 \ \Omega.$$

Figure 10.11 shows the circuit.

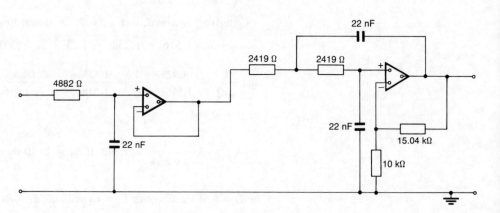

Fig. 10.11

High-pass Filters

The low-pass Sallen and Key filter shown in Fig. 10.9 can be converted into a high-pass filter by interchanging the resistors with the capacitors. Figure 10.12 shows the circuit of an active high-pass filter.

The design of a high-pass filter commences with the transformation of the nth-order low-pass transfer function into the corresponding high-pass function. Consider the first-order low-pass filter; its transfer function is $F(S) = 1/(1+SCR) = 1/(1+S/\omega_0)$. For the first-order high-pass filter shown in Fig. 10.13,

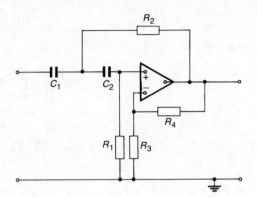

Fig. 10.12 High-pass active filter.

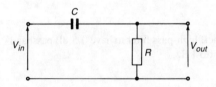

Fig. 10.13 First-order high-pass filter.

$$F(S) = \frac{V_{out}(S)}{V_{in}(S)} = \frac{R}{1+1/SC} = \frac{SCR}{1+SCR} = \frac{S/\omega_0}{1+S/\omega_0}.$$

If S/ω_0 is replaced by ω_0/S^1, $F(S)$ becomes

$$\frac{\omega_0/S^1}{1+\omega_0/S^1} = \frac{1}{1+S^1/\omega_0}$$

which is identical to $F(S)$ for the low-pass filter.

This means that to convert a low-pass filter's transfer function to a high-pass one, write ω_0/S for S/ω_0.

Thus, the general expression for a low-pass filter

$$\frac{H_0}{\dfrac{S^2}{\omega_0^2} + \dfrac{S}{\omega_0 Q} + 1} \quad \text{becomes} \quad \frac{H_0}{\dfrac{\omega_0^2}{S^2} + \dfrac{\omega_0}{SQ} + 1}$$

$$\text{or} \quad \frac{S^2 H_0}{\omega_0^2 + \dfrac{S\omega_0}{Q} + S^2}$$

which is the general expression for a high-pass filter.

For a Butterworth filter no values need be altered to change from low-pass to high-pass; it is merely necessary to interchange the positions of the resistors with those of the capacitors. The design of a Tchebysheff high-pass filter is somewhat more complicated.

Example 10.8

Design a Butterworth second-order high-pass filter to have a cut-off frequency of 3 kHz.

Solution

From Example 10.5, $A_{v(f)} = 1.586$, $R_3 = 10$ kΩ, $R_4 = 5.86$ kΩ, $C_1 = C_2 = 22$ nF and $R_1 = R_2 = 2410$ Ω. The circuit of the filter is given by Fig. 10.14.

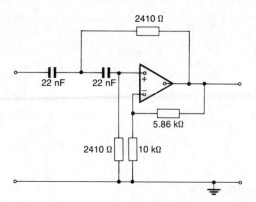

Fig. 10.14

Example 10.9

Design a second-order Tchebysheff high-pass filter to have 0.5 dB passband ripple and a ripple bandwidth of 3 kHz.

Solution
From Table 10.3

$$F(S) = \frac{H_0}{S^2 + 1.425S + 1.516} = \frac{H_0}{\dfrac{S^2}{1.516\omega_r^2} + \dfrac{1.425S}{1.516\omega_r} + 1}$$

$$= \frac{H_0 S^2}{\dfrac{\omega_r^2}{1.516} + \dfrac{1.425S\omega_r}{1.516} + S^2}.$$

Comparison with equation (10.4) gives

$$\omega_0 = \frac{\omega_r}{\sqrt{1.516}} = \frac{2\pi \times 3000}{\sqrt{1.516}} = 15\ 309 \text{ rad/s}$$

and $\dfrac{\omega_0}{Q} = \dfrac{1.425\omega_r}{1.516}$ or $Q = 0.864.$

Hence $A_{v(f)} = 1.843$, $\omega_0 = \dfrac{1}{RC} = 15\ 309.$

Choosing $C_1 = C_2 = 22$ nF gives

$$R_1 = R_2 = R = \frac{10^9}{15\ 309 \times 22} = 2969\ \Omega.$$

The State Variable Active Filter

In the design of an active filter the concern is for how close the produced transfer function is to the desired response. The reason for some deviation between the object and the obtained responses are the tolerances of the resistors and the capacitors, as well as the long-term stability of the op-amp. Any changes in the op-gain or bandwidth may affect the response of the filter.

The *sensitivity* of the filter is a measure of the change in its transfer function with respect to a change in a component value, or to a change in op-amp gain. The Sallen and Key type of filter has a rather high sensitivity and may therefore not always be suitable for all applications. A better filter, from the sensitivity point of view, is the state variable filter shown in Fig. 10.15.

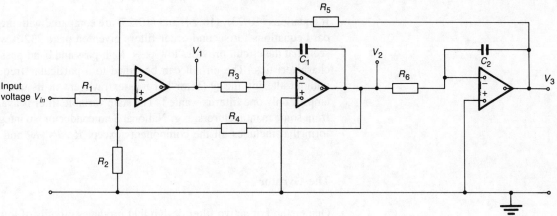

Fig. 10.15 State variable filter.

For this circuit, assuming that $R_3 = R_6 = R$ and $C_1 = C_2 = C$.

$$V_1 = V_{in} + V_2 - V_3 \qquad (10.20)$$
$$V_2 = -V_1/RCS \qquad (10.21)$$
$$\text{and} \quad V_3 = -V_2/RCS = V_1/R^2C^2S^2. \qquad (10.22)$$
$$\text{Therefore, } V_1 = V_{in} - V_1/RCS - V_1/R^2C^2S^2$$

$$V_{in} = V_1\left(1 + \frac{1}{RCS} + \frac{1}{R^2C^2S^2}\right)$$

$$\text{or} \quad V_1 = \frac{S^2 V_{in}}{S^2 + \dfrac{S}{RC} + \dfrac{1}{R^2C^2}}. \qquad (10.23)$$

Rewriting equation (10.20) in terms of V_2,

$$-RCSV_2 = V_{in} + V_2 + \frac{V_2}{RCS}$$

$$V_{in} = -V_2\left(1 + RCS + \frac{1}{RCS}\right)$$

or $\quad V_2 = \dfrac{SV_{in}/RC}{S^2 + \dfrac{S}{RC} + \dfrac{1}{R^2C^2}}.$ (10.24)

Rewriting equation (10.20) in terms of V_3

$$R^2C^2S^2V_3 = V_{in} - RCSV_3 - V_3$$
$$V_{in} = V_3(1 + RCS + R^2C^2S^2)$$

or $\quad V_3 = \dfrac{V_{in}/R^2C^2}{S^2 + \dfrac{S}{RC} + \dfrac{1}{R^2C^2}}.$ (10.25)

If equations (10.23), (10.24) and (10.25) are compared with the standard equations for second-order filters given on page 232 it will be seen that the circuit provides low-pass, high-pass and band-pass filter characteristics. The circuit can be tuned to a particular frequency without altering the Q-factor but it is extravagent in its use of op-amps if only one filter is wanted. The state variable filter is available from some manufacturers, e.g. National Semiconductor, in integrated form that includes all the components except R_1, R_2, R_3 and R_6.

The Gyrator

One method of active filter design that produces circuits of low sensitivity consists of simulating the inductors in well-tried traditional LC designs. The simulation of the inductances is achieved by means of a circuit known as a *gyrator*. The gyrator is a circuit which, if a capacitor is connected across its output terminals will generate an effective inductance at its input terminals. The idea is shown by Fig. 10.16, g_1 and g_2 are the *gyration conductances* of the gyrator. The gyrator can be obtained as an IC but it is probably more often produced using two, or three, op-amps.

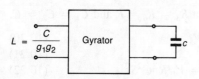

Fig. 10.16 The gyrator.

11 Power Supplies

Modern electronic, computer, radio and telecommunication equipment depends for its correct operation upon the availability of stable, well-regulated d.c. power supplies. The main requirements for a power supply are good regulation, low noise and ripple, low output impedance, and output short-circuit protection. The block diagram of a mains power supply is shown by Fig. 11.1; it contains the following sections: transformer, rectifier, filter, and voltage regulator. The transformer changes the mains supply voltage into the required lower voltage that is somewhat more than the wanted d.c. output voltage. The transformed voltage is then rectified to give a fluctuating d.c. voltage and then this is passed through the filter. The filter is required to attenuate the *ripple* component of the rectified voltage; in most modern power-supply circuits the filter is merely a single shunt capacitor since the following voltage regulator effectively rejects any ripple present at its input terminals. The function of the voltage regulator is to provide a constant d.c. output voltage even though the input a.c. voltage and/or the load current may vary.

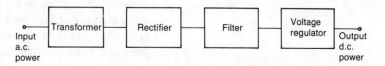

Fig. 11.1 Block diagram of a power supply.

Full-wave Rectification

Full-wave rectification is normally employed using either of the two circuits shown in Fig. 11.2, both of which include the capacitor filter. The bridge rectifier, Fig. 11.2(*b*), uses four diodes instead of two, but it does not demand the use of a transformer with a centre-tapped secondary winding. Bridge rectifiers are sold as complete packages. The waveforms of the output voltage from the rectifier itself, the capacitor voltage, and the capacitor current are shown by Fig. 11.3. It can be seen that the capacitor has the effect of *smoothing* the output voltage. While the diode is conducting, the secondary voltage V_s of the transformer is applied directly across the load (minus the small

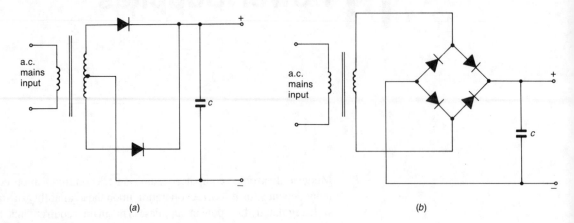

Fig. 11.2 (a) Full-wave rectifier and (b) bridge rectifier.

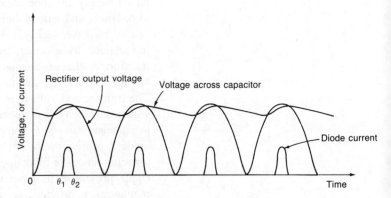

Fig. 11.3 Full-wave rectifier current and voltage waveforms.

diode voltage drop), and so the output voltage is $v_0 = V_S \sin \omega t$. The capacitor charges up to the peak value V_S of the secondary voltage. After the peak of the secondary voltage has passed, v_0 falls towards zero volts and is very quickly smaller than the voltage across the capacitor. Immediately this happens the diode turns off. The capacitor then discharges, at a rate determined by the time constant CR_L seconds, through the load resistance R_L. The diode will remain non-conducting until the secondary voltage, on its next positive half-cycle, again becomes larger than the voltage across the capacitor. The diode will then turn on and conduct current; this means that the diode currents flow in a series of narrow pulses as shown. The times at which the diode current commences, and then stops, flowing are known as the *cut-in* θ_1, and the *cut-out* θ_2, *points*, respectively.

Output Voltage

An approximate, but reasonably accurate, expression for the d.c. output voltage of a full-wave rectifier can be derived if it is assumed that the time constant CR_L is large enough for the fall of the capacitor

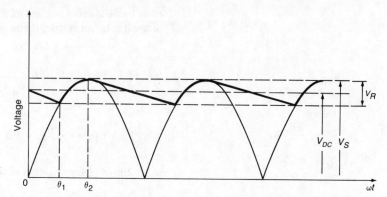

Fig. 11.4 Calculation of d.c. output voltage.

voltage to be taken as being linear instead of exponential. Consider Fig. 11.4. Assuming the capacitor to be of large value so that $\omega CR_L \gg 1$, then the cut-in point θ_1 is very nearly $\pi/2$ rad, when the output voltage is at its maximum value of V_S volts. If the change in the capacitor voltage is V_R volts — which is the peak—peak ripple voltage — then Fig. 11.4 shows that the average, or d.c., value of the output voltage is

$$V_{DC} = V_S - V_R/2. \tag{11.1}$$

The time per half-cycle for which zero diode current flows is t seconds; in this time the capacitor loses a charge of $I_{DC}t$ coulombs. Since $V = Q/C$,

$$V_R = \frac{I_{DC}t}{C}. \tag{11.2}$$

Since the diode current pulses are narrow the time t is very nearly equal to one-half of the periodic time T of the a.c. input voltage. Hence $t \simeq T/2 = 1/2f$, where f is the frequency of the a.c. voltage. Therefore,

$$V_R = \frac{I_{DC}}{2fC}. \tag{11.3}$$

Substituting equation (11.3) into equation (11.1),

$$V_{DC} = V_S - \frac{I_{DC}}{4fC}. \tag{11.4}$$

For the output d.c. voltage to have a small ripple content a large value of capacitance is necessary and so an electrolytic type is always employed.

Example 11.1

A full-wave rectifier operates from a 50 Hz mains supply. The voltage across the secondary winding of the transformer is 35 V peak. If the mean current taken by the load is 4 mA and a 200 μF smoothing capacitor is used, calculate (a) the d.c. output voltage, (b) the peak ripple voltage, and (c) the ripple factor.

Solution
(a) From equation (11.4),

$$V_{DC} = 35 - \frac{4 \times 10^{-3}}{4 \times 50 \times 200 \times 10^{-6}} = 34.9 \text{ V. } (Ans.)$$

(b) Peak ripple voltage = 0.1 V.   (*Ans.*)

(c) Ripple factor γ is defined as $\dfrac{\text{r.m.s. ripple voltage}}{\text{d.c. output voltage}}$ and hence

$$\gamma = \frac{0.1}{\sqrt{3} \times 34.8} = 0.165\%. \quad (Ans.)$$

($\sqrt{3}$ appears since the ripple voltage is of approximately triangular waveshape.) This value for the ripple factor is on the high side; for most applications $\gamma \leq 0.1\%$ is wanted.

Example 11.2

A full-wave rectifier circuit supplies a load of 2000 Ω resistance with a d.c. current of 8 mA. If the peak−peak ripple voltage is to be 0.5 V calculate the minimum possible value for the smoothing capacitor. The mains frequency is 50 Hz.

Solution

$$\frac{I_{DC}t}{C} = \Delta V, \ \frac{8 \times 10^{-3}}{100 \ C} = 0.5$$

or $C = 160 \ \mu$F.   (*Ans.*)

Voltage Regulation

The voltage regulation of a power supply is the variation of the d.c. output voltage with change in either the load current, or in the input voltage.

(a) The *load regulation* is the maximum change in the d.c. output voltage as the load current is varied from its minimum (usually no-load) value to its maximum rated load. The input voltage is assumed to be constant at its nominal value. The percentage load regulation is defined by different manufacturers in either one of two different ways:

(i) percentage regulation $= \dfrac{V_{\text{no-load}} - V_{\text{full-load}}}{V_{\text{full-load}}} \times 100,$

$$(11.5(a))$$

$$(ii) \text{ percentage regulation} = \frac{V_{\text{no-load}} - V_{\text{full-load}}}{V_{\text{no-load}}} \times 100.$$

$$(11.5(b))$$

(b) The *line regulation* is the maximum change in the d.c. output voltage as the input voltage is varied over its possible range of values. The load current is assumed to remain constant.

To improve the regulation and the ripple rejection of a power supply some form of voltage regulator is nearly always used. There are two main classes of regulator, namely: linear regulators and switched-mode regulators. The former is widely employed for medium-power supplies and the latter for supplies that must output a large current at low or medium voltage.

Linear Voltage Regulators

The block diagram of a linear voltage regulator is shown by Fig. 11.5. The output voltage of the rectifier is applied to a series-control element which introduces resistance into the positive supply line. The output voltage V_{OUT} is smaller than the input voltage by the voltage dropped across the series element. The output voltage, or a known fraction of it, is compared in the voltage comparator with a voltage reference. The difference between the two voltages is detected and an amplified version of it is applied to the series-control element in order to vary its resistance in such a way as to maintain the output voltage at its correct value. If, for example, the output voltage is larger than it should be, the amplified difference voltage will be of such a polarity that the resistance of the controlled element will be made larger and the output voltage will fall. Conversely, if the output voltage is less than its correct value, the resistance of the series-controlled element will be reduced by the amount necessary for the output voltage to rise to its correct value. Generally, the series-control element is a transistor connected as shown in Fig. 11.6. When an n-p-n transistor is employed, its collector is connected to the input terminal, and its emitter is connected to the output terminal, of the circuit since the former is more positive. If the output voltage of the stabilizer should vary by an amount δV_{OUT} the control voltage appearing at the base terminal of the series transistor will be $A_v' \delta V_{OUT}$, which is the

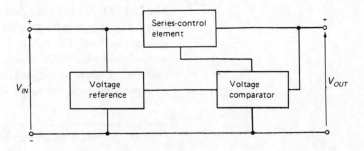

Fig. 11.5 Block diagram.

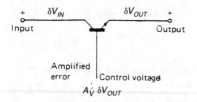

Fig. 11.6 Transistor series-control element.

amplified error voltage produced by the comparator. The base-emitter voltage of the transistor is then

$$A_v' \delta V_{OUT} - \delta V_{OUT}$$

and will produce a voltage

$$A_v'' (A_v' \delta V_{OUT} - \delta V_{OUT})$$

between the base and collector terminals where A_v'' is the voltage gain of the series transistor. The voltage across this transistor is also equal to $\delta V_{IN} - \delta V_{OUT}$ and therefore

$$\delta V_{IN} - \delta V_{OUT} = A_v'' (A_v' \delta V_{OUT} - \delta V_{OUT}). \tag{11.6}$$

Example 11.3

In a voltage stabilizer of the type shown in Fig. 11.5 the error voltage gain of the comparator is -120 and the voltage gain of the series transistor is -20. The rectifier circuit connected to the input terminals of the stabilizer has an output resistance of $100\ \Omega$. Calculate the change in the output voltage that occurs when the load current changes by 10 mA.

Solution

$$\delta V_{IN} = \delta I_L R_{OUT} = 10 \times 10^{-3} \times 100 = 1\ \text{V}.$$

Hence, substituting into equation (11.6),

$$1 - \delta V_{OUT} = -20(-120\delta V_{OUT} - \delta V_{OUT})$$
$$1 = 2400\delta V_{OUT} + 20\delta V_{OUT} + \delta V_{OUT}$$

$$\delta V_{OUT} = \frac{1}{2421} = 0.413\ \text{mV}. \quad (Ans.)$$

Under the worst-case conditions of high input voltage and low output voltage the power dissipated in the series-control element may become excessive. The efficiency of a linear regulator is usually about 25% but may be as low as 15% under the worst-case condition.

Modern electronic equipments generally employ an integrated circuit voltage regulator for the usual IC reasons. Indeed, there is often one main power supply with little, if any, regulation which supplies a number of circuit boards each of which has its own on-board IC voltage regulator. A wide variety of IC regulators are available to the designer and here only one family of devices will be considered.†
The LM 117/317 ICs are a range of three-terminal positive-variable-output, voltage regulators which are able to supply maximum load currents of from 100 mA to 1.5 A. If a negative output voltage is wanted the LM 137/337 can be used instead. The LM 317 can, for example, produce an output voltage in the range 1.2 V to 37 V with an input voltage in the range 4 to 40 V; but the maximum voltage

† The 723 and 78xx series are discussed in *Electronics II*.

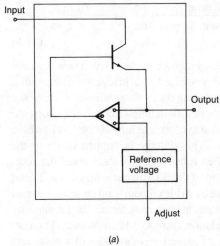

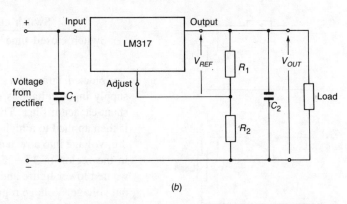

(b)

Fig. 11.7 *(a)* Internal block diagram of LM 317 voltage regulator. *(b)* LM 317 connected in a power supply.

differential between input and output must not be greater than 40 V. The load regulation is 0.1% and the line regulation is 0.01%. The ripple rejection is 65 dB. All the ICs in this group incorporate current-limit, thermal-overload, and safe-area shut-down features.

Figure 11.7(*a*) shows the basic internal block diagram of a LM 317 and Fig. 11.7(*b*) shows how the device is connected at the output of a full-wave rectifier. The d.c. output voltage V_{OUT} of the regulator is set by the choice of the resistor values R_1 and R_2. The reference voltage is nominally 1.25 V but it may vary within the limits 1.20 V to 1.30 V. The reference voltage appears between the output and adjust terminals of the IC and hence also across resistor R_1. As a result, a constant current equal to V_{REF}/R_1 flows through resistor R_2. The output voltage V_{OUT} of the circuit is

$$V_{OUT} = V_{REF} + \frac{V_{REF}R_2}{R_1} = V_{REF}\left(1 + \frac{R_2}{R_1}\right). \qquad (11.7)$$

There is, in addition, a small voltage of $I_{ADJ}R_2$, but since $I_{ADJ} \simeq 50\ \mu\text{A}$, this component is negligibly small.

It is usual to make $R_1 = 240\ \Omega$ and to select R_2 to give the wanted output voltage using the nominal 1.25 V value for V_{REF}. The regulator must have a voltage drop of at least 3 V across it or it will not work. Also, the safe area [the product $(V_{IN} - V_{OUT})I_{DC}$], must not exceed the rated maximum power dissipation. In both cases, once the excess condition is removed the IC will resume its normal operation. The capacitor C_1 is required if the leads from the rectifier are of considerable length, while C_2 improves the transient response. Typically, $C_1 = 0.1\ \mu\text{F}$ and $C_2 = 1\ \mu\text{F}$.

Switched-mode Power Supplies

The switched-mode power supply (s.m.p.s.) is often used because it has a much higher efficiency than a linear supply (typically 65% to 90%) and, because of this, it is both smaller and lighter in weight. Unlike a linear regulator, which uses a variable linear series element to control the load current (see Fig. 11.8(*a*)), a s.m.p.s. uses a series switch that is either open or closed (see Fig. 11.8(*b*)). The d.c. output voltage is proportional to the ratio

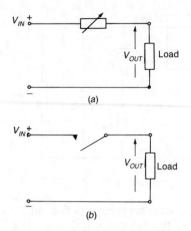

Fig. 11.8 Principle of operation of (a) linear series voltage regulator and (b) switched-mode voltage regulator.

$$\frac{\text{Switch closed time}}{\text{Switch closed time} + \text{switch open time}} = \frac{t_{on}}{t_{on} + t_{off}}.$$

(11.8)

The block diagram of a s.m.p.s. is shown in Fig. 11.9. The a.c. mains supply input voltage is directly applied to a bridge rectifier with a shunt-capacitor filter. The fluctuating d.c. voltage across the capacitor is then applied to a high-frequency switching circuit. This chops the d.c. voltage into a rectangular waveform, at a frequency somewhere in the region of 15–25 kHz. This voltage is stepped down to the wanted lower figure and is then filtered to give the wanted d.c. output voltage. Voltage regulation of the d.c. output voltage is achieved by means of the control and pulse-width modulation (p.w.m.) circuit which closes the loop from output to switching circuit. In the majority of circuits pulse-width modulation (p.w.m.) is employed. The d.c. output voltage is equal to tV_{IN}/T, where t is the width of a pulse and T is the periodic time of the pulse waveform, and so the voltage across the load is determined by the ratio t/T. If the output voltage of the s.m.p.s. should rise (because either the a.c. input voltage has increased or the load has altered) the control circuit will apply narrower pulses to the switching circuit. Conversely, if the d.c. output voltage should fall, wider pulses will be applied.

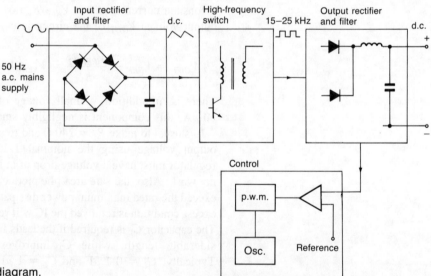

Fig. 11.9 S.m.p.s. block diagram.

There are two main kinds of s.m.p.s., known respectively as the *forward converter* and the *flywheel converter*. Figure 11.10 shows the basic circuit of a forward-converter s.m.p.s. The input a.c. voltage is rectified and the d.c. voltage produced is applied to the collector of transistor T_1 via the primary winding of the transformer. The voltage which appears across L_1 is stepped down by the transformer

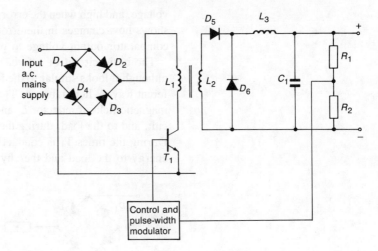

Fig. 11.10 Forward-converter s.m.p.s.

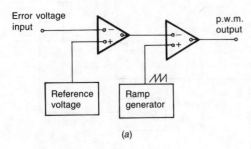

(a)

Fig. 11.11 Control and p.w.m. circuit (a) block diagram and (b) waveforms.

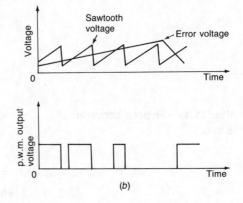

(b)

action and rectified by diode D_5. L_3 and C_1 act as a low-pass filter to smooth the rectified voltage and produce the output d.c. voltage. When T_1 is off the energy stored in inductor L_2 is passed to the load via the *free-wheeling diode* D_6. The filtering action of L_3C_1 is very efficient because of the high frequency of the secondary rectangular voltage. A fraction of the output voltage is fed back to the control circuit and the pulse-width modulator. Here the error voltage is translated into a variable duty cycle pulse train at a fixed frequency (set by an oscillator). The output voltage is thus regulated in the manner described earlier. The basic block diagram of the control and p.w.m. IC is shown by Fig. 11.11(a) and the circuit waveforms by Fig. 11.11(b). The frequency of the ramp waveform is set by an external capacitor connected across the appropriate terminals. The difference between the fraction of the output voltage fed back and the internal reference voltage is amplified to produce an error voltage that varies with time. The error voltage is compared with the output of a ramp generator in another comparator. The output of this comparator is low whenever the ramp voltage is bigger than the error voltage, and high when the error voltage is the larger. Figure 11.11(*b*) shows how changes in the error voltage vary the duty cycle of the comparator output voltage to produce a p.w.m. pulse train.

The basic circuit of a flyback converter is given in Fig. 11.12; although it looks similar to the forward converter it operates in a different way. When transistor T_1 is conducting, energy is stored in the magnetic field of inductor L_1 and this is transferred to the output circuit, and to the load, during the flyback period, i.e. when T_1 is off. During the times T_1 is conducting capacitor C_1 continues to supply energy to the load and thereby gives a smoothing action.

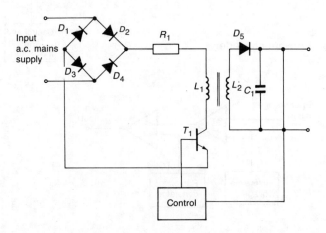

Fig. 11.12 Flyback converter s.m.p.s.

The ripple content of the output voltage of the flyback converter is higher than for the forward converter but no series output inductor is needed. In practice, flyback converters are used for powers of 50 to 100 W and forward converters for powers of 100 to 200 W. Other types of converter are used for even higher powers.

Merits of Linear and Switching Regulators

The advantages of a s.m.p.s. are given below.

(*a*) since the transformer operates at a relatively high frequency a much smaller and cheaper transformer can be used.
(*b*) Smoothing of the input voltage is carried out at 320 V/100 Hz instead of the lower voltage used in a linear circuit. This reduces the size of the smoothing capacitor.
(*c*) Smoothing the output voltage is much easier since the ripple frequency is 15 kHz or more.
(*d*) The voltage regulation is better.
(*e*) The efficiency is higher, typically about 85%.

The disadvantages of the s.m.p.s. are:

(a) radio-frequency interference (r.f.i.) is generated and careful design is necessary to keep this within acceptable limits;

(b) the transient response is not as good as that of a linear circuit; and

(c) the output noise and ripple content of a linear regulator is smaller.

Current Limiting and Remote Sensing

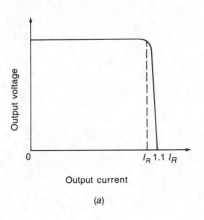

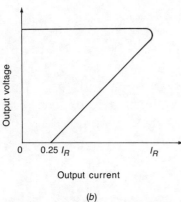

Fig. 11.13 Current/voltage characteristic of (a) current limiter, (b) foldback current limiter.

Current Limiting

There is always the possibility that a power supply may inadvertently be overloaded or short-circuited and steps must be taken to ensure that no damage is then caused. All but the simplest power supplies therefore include some kind of *current limiting*. A current-limiting circuit monitors the current supplied to the load, and if this should exceed its rated value by a specified amount, it automatically reduces the current. The current/voltage characteristic of a current limiter is shown by Fig. 11.13(a). If the normal load current is fairly high a large heat sink will be necessary to dissipate the heat generated within the supply by the short-circuit current. A more efficient arrangement is known as *foldback over-current protection*. With this, if the load current becomes excessively high, *both* the output current *and* the output voltage are reduced (see Fig. 11.13(b)). Foldback current limiting requires more complex circuitry but it gives a much lower internal power dissipation under short-circuit conditions. It is clear from Fig. 11.13 that for an output voltage of zero (short-circuit load) the current is, typically, 1.1 times the rated maximum for normal limiting, but only 0.25 times the rated maximum current for foldback limiting.

Remote Sensing

Some power supplies incorporate a feature known as *remote sensing*. Remote sensing of a load voltage is illustrated by Fig. 11.14. The power supply has two sensing lines which are additional to the load-current carrying wires. The sensing wires allow the power supply to achieve optimum regulation at the load rather than at the power supply output terminals. This compensates for the voltage drop in the leads taking the current to the load.

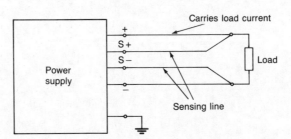

Fig. 11.14 Remote sensing.

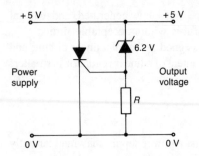

Fig. 11.15 Crowbar over-voltage protection.

Over-voltage Protection

This is a feature of some power supplies that shuts down the supply if the output voltage rises above a pre-determined level. Figure 11.15 shows a *crowbar* voltage-protection circuit; if the output voltage rises to a value in excess of 6.2 V the Zener diode conducts and the resulting voltage across the resistor triggers the thyristor ON. This places a near short-circuit across the output terminals and so shuts down the system.

Exercises

Exercises 1

1.1 A power transistor has a maximum collector dissipation of 4 watts for an ambient temperature of 30 °C. If the maximum junction temperature is 170 °C and the thermal resistance θ_{JS} between junction and heat sink is 4 °C/W, calculate the maximum possible thermal resistance for the heat sink.

1.2 A transistor is to dissipate 12 W power at an ambient temperature of 45 °C with a maximum collector temperature of 90 °C. The thermal resistance between junction and case is 0.9 °C/W and between case and air is 6 °C/W. Calculate the thermal resistance of the required heat sink. Find also the maximum power which can be dissipated if a mica washer of thermal resistance 1.8 °C/W is inserted between the case of the transistor and the heat sink.

1.3 Define the four h-parameters and draw the h-parameter equivalent circuit of a bipolar transistor having a resistive load R_L. A transistor has the following parameters: h_{ie} = 600 Ω, h_{fe} = 100, h_{oe} = 20 × 10^{-6} S, and h_{re} = 2 × 10^{-4}. Calculate the voltage gain of the transistor when R_L = 3.3 kΩ. Determine also the percentage error involved in neglecting (a) h_{re} only, (b) h_{re} and h_{oe}.

1.4 Define each of the four y-parameters and draw the y-parameter equivalent circuit of a bipolar transistor having a resistive load R_L. A common-emitter transistor has a 7.8 kΩ load and the following y-parameters: y_{ie} = 500 × 10^{-6} S, y_{oe} = 50 × 10^{-6} S, y_{fe} = 0.025 S, y_{re} = 0. Calculate (a) the input and output impedances, (b) the voltage gain of the circuit.

1.5 Explain the factors that cause the current gain of a bipolar transistor to fall at high frequencies.

A transistor has a current gain h_{fe} of 100 at audio frequencies. The 3 dB cut-off frequency is 400 kHz. Calculate the current gain to be expected at (a) 800 kHz, (b) 2 MHz, (c) 4 MHz.

1.6 What is meant by the Miller effect? Derive an expression for the input capacitance of a field-effect transistor.

A FET has C_{gs} = 5 pF, C_{gd} = 1.5 pF and g_m = 2.4 mS and a drain load resistor of 4.7 kΩ. Calculate the high frequency at which the voltage gain of the circuit has fallen by 3 dB on its low-frequency value if the source resistance is 1000 Ω.

1.7 A bipolar transistor has $r_{bb'}$ = 80 Ω, $r_{b'e}$ = 700 Ω, $C_{b'e}$ = 50 pF and $C_{b'c}$ = 5 pF, and g_m = 40 mS. Calculate the value of collector load resistor to give the voltage gain an upper 3 dB frequency of 5 MHz. Find also the low-frequency gain.

1.8 The specifications of a transistor include:

$V_{BE(SAT)}$	I_C = 0.5 A	I_B = 50 mA	1.2 V_{max}
	I_C = 2.0 A	I_B = 200 mA	1.5 V_{max}
C_{TC}	V_{CB} = 5.0 V	I_E = 0	60 pF
f_t	I_C = 0.25 A	V_{CE} = 5.0 V	T_{AMB}
		= 25 °C	60 MHz mi
h_{FE}	I_C = 0.5 A	V_{CE} = 12 V	78 min
			250 max
	I_C = 2.0 A	V_{CE} = 10 V	40 min
$V_{CBO(MAX)}$	I_C ≤ 1.0 mA		70 V
$V_{CEO(MAX)}$			45 V
$I_{C(MAX)}$			6.0 A
P_{TOTAL}	T_{AMB} ≤ 60 °C		11 W

Explain fully the meaning and importance of each parameter. What factors limit the maximum power that can be dissipated in a transistor?

1.9 A transistor specification includes the following data:

maximum operating junction temperature	175 °C
maximum dissipation at 25 °C case temperature	30 W
maximum dissipation at 25 °C ambient temperature	12 W

From these data sketch a typical power—temperature derating curve.

Determine the thermal resistance of a heat sink for this transistor if it is to dissipate 20 W in an ambient temperature of 60 °C. What is the resulting case temperature? Assume that the thermal resistance between the case and the heat sink is 2 °C/W. What practical steps can be taken to minimize this resistance?

1.10

(a) Draw the *h*-parameter equivalent circuit of a transistor and define each parameter. State the limitations on its use.

(b) Draw the hybrid-π equivalent circuit of a bipolar transistor and relate each resistance to the parameters of the transistor.

(c) A bipolar transistor has $h_{ie} = 1100$ Ω, $h_{fe} = 180$ and $h_{oe} = 80$ μS. Calculate values for the hybrid-π equivalent circuit components.

1.11 An amplifier has input impedance of $1000 \angle 45°$ Ω, an output impedence of $50 \angle 10°$ Ω, and a current gain of $150° \angle 0°$ at a particular frequency. Calculate its voltage gain when the load impedance is $200 \angle 0°$ Ω.

1.12

(a) The f_t of a bipolar transistor is 500 MHz. Calculate its current gain at (a) 100 MHz, (b) 50 MHz.

(b) A transistor has $f_t = 600$ MHz and $g_m = 30$ mS. Determine the sum of its capacitances $C_{b'e}$ and $C_{b'c}$.

1.13 Calculate the current gain, input resistance and voltage gain of a Darlington pair if, for T_1, $h_{ie} = 1000$ Ω, $h_{fe} = 120$, and, for T_2, $h_{ie} = 300$ Ω, $h_{fe} = 80$. The load resistance is 1 kΩ.

Exercises 2

2.1 A common-emitter transistor has a collector load of 5 kΩ and is coupled by a large capacitor to another stage having an input impedance of 2.5 kΩ. Draw the a.c. equivalent circuit of the first stage at high frequencies and calculate (a) the gain–bandwidth product, (b) the frequency at which the gain has fallen by 10 dB on its mid-frequency value, if the mid-frequency gain is 120 and the total shunt capacitance across the load is 100 pF.

2.2 A common-emitter transistor has a collector load of 5 kΩ and is coupled by a large capacitor to an external load of 2.5 kΩ. If the total shunt capacitance across the load is 200 pF calculate the relative voltage gain of the circuit at 1.2 MHz. Derive any expression used.

2.3 Draw the *y*-parameter equivalent circuit of the amplifier shown in Fig. Q.1. Use the circuit to calculate (a) the input impedance, (b) the mid-frequency voltage gain, and (c) the upper 3 dB frequency of the amplifier, given $y_{ie} = 3 \times 10^{-3}$ S, $y_{fe} = 200 \times 10^{-3}$ S, $y_{re} = y_{oe} = 0$.

2.4 Discuss the reasons for the fall-off in the voltage gain of an audio amplifier at low frequencies.

An amplifier has a collector load resistor of 4.7 kΩ and is coupled to an external load resistance of 2 kΩ by a

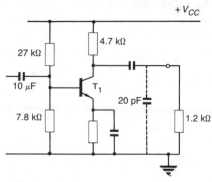

Fig. Q.1

capacitor. Calculate the value of this coupling capacitor if the lower 3 dB frequency is to be 50 Hz. Calculate also the gain, relative to mid-frequencies, at 25 Hz.

2.5 A common-emitter transistor is driven by a signal source of e.m.f. 10 μV and output resistance 2 kΩ. If the effective collector load resistance is 4 kΩ calculate (a) the current gain, (b) the input resistance, (c) the voltage gain, and (d) the output resistance of the circuit. The *h*-parameters of the transistor are $h_{ie} = 4800$ Ω, $h_{fe} = 300$ and $h_{oe} = 22$ μS.

2.6 A single-stage transistor amplifier has an input resistance of 1000 Ω, an upper 3 dB frequency of 157 kHz, and a mid-frequency voltage gain of 200 when the collector load resistance is 2 kΩ and the impedance of the signal source is 1000 Ω. When the collector load resistance is reduced to 200 Ω the upper of 3 dB frequency is 1 MHz and the mid-frequency gain is 20. Calculate the transistor capacitances $C_{b'e}$ and $C_{b'c}$.

2.7 Show that the c.m.r.r. of a differential amplifier is given by $1 + 2R_E(1 + h_{fe})/h_{ie}$, where R_E is the emitter resistance. Calculate the necessary value of R_E to obtain a c.m.r.r. of 90 dB if $h_{fe} = 120$ and $h_{ie} = 1200$ Ω.

2.8 A potential divider bias circuit has been designed for use with a bipolar transistor having a nominal h_{FE} of 80. The effective base resistance $R_1R_2/(R_1+R_2)$ is 6 kΩ and the collector current is 1.5 mA. Determine the minimum value of emitter resistance needed to limit the collector current to 1.55 mA when a transistor having the maximum h_{FE} of 100 is used.

2.9 Design a potential divider bias circuit for an a.f. transistor stage if $I_C = 1$ mA, $V_{CC} = 12$ V, $V_{CE} = 5.5$ V, $h_{FE} = 120$ and $V_{BE} = 0.6$ V. Calculate the values of I_C and V_{BE} if h_{FE} increases to 150.

2.10 The LM 381 circuit of Fig. 2.12(b) is to have a voltage gain of 500 and a lower 3 dB frequency of 25 Hz. Calculate suitable values for R_1, R_2, R_3, and C_1.

2.11 The results obtained from an amplifier test are given in Table Q.1.

Table Q.1

Frequency (kHz)	0.01	0.02	0.06	0.1	0.2	0.6	1.0	3.0	5.0	10.0
Gain (dB)	40	39.5	34	27	17	−3	−14	−36	−46	−60
Phase lag (°)	100	112	125	132	143	162	170	185	193	205

Plot the Bode diagrams using a common axis for both gain and phase angle and determine:

(a) the gain margin,
(b) the phase margin,
(c) the gain crossover frequency,
(d) the phase crossover frequency, and
(e) the frequency at which the system will oscillate if the gain is steadily increased. State the important assumption made in this case.

2.12 A transistor has a collector load which consists of a 318 μH inductor of Q factor 100, in parallel with a capacitor. The output capacitance and conductance of the transistor are 9 pF and 5 μS, respectively. Calculate the capacitance value and the Q factor. The amplifier is now connected to the next stage via a transformer. This stage has an input conductance of 300 μS and negligible capacitance. The turns ratio of the transformer is 7:1 step-down. Determine the effective Q factor and 3 dB bandwidth of the amplifier, $f_o = 464.6$ kHz.

Exercises 3

3.1 An amplifier has a voltage gain of

$$A_v = \frac{-1000}{(1+j10^{-5}f)^3}$$

before feedback is applied. The feedback factor is 1/250. Plot the Nyquist diagram and use it to confirm that the amplifier is stable. Determine the gain margin of the amplifier.

3.2 Figure Q.2 shows the Bode amplitude and phase plots of an amplifier. Determine the feedback factor required for a phase margin of (a) 30°, (b) 60°.

3.3 Draw the circuit of an emitter follower and explain why bootstrap biasing is sometimes used.

An emitter follower is connected between a 1000 Ω source and a 100 Ω load. If the follower is to be matched to the load, determine the possible values for h_{fe} and the emitter

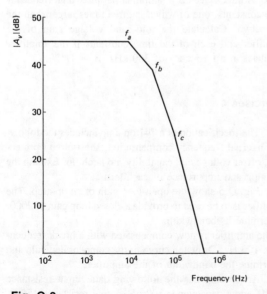

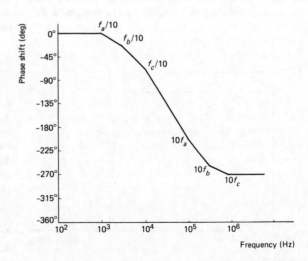

Fig. Q.2

resistance R_E. What is then the input resistance of the circuit?

3.4 A three-stage n.f.b. amplifier has a gain of $A_{v(f)} = 50$. If the gain of each stage falls by 10% the overall gain $A_{vo(f)}$ is required to fall by only 0.5%. Calculate the necessary overall gain A_v and the feedback factor of the amplifier before n.f.b. is applied.

3.5 Discuss the reduction of non-linearity distortion in an n.f.b. amplifier.

Before n.f.b. is applied, an amplifier has an output power of 5 W in a load of 12 Ω with 10% harmonic distortion. Calculate the necessary value of βA for the percentage distortion to be reduced to 0.5%. What assumptions are made?

3.6 An amplifier has voltage-series negative feedback applied to it. If the gain before feedback is $460 \angle 0°$ at 2 kHz and $230 \angle -60°$ at 20 kHz and $\beta = 0.01$ calculate the gain at (a) 2 kHz, (b) 20 kHz.

3.7 The specification of an integrated-circuit amplifier includes (a) nominal gain 80 dB, (b) maximum gain variation 75−85 dB, (c) input impedance 120 kΩ. The IC is used with voltage-series n.f.b. to reduce the gain to 60 dB. Calculate (a) the maximum gain variation, (b) the input impedance of the feedback amplifier.

3.8 Explain why an amplifier may become unstable at either a low or a high frequency. How many time constants are necessary for instability?

A particular amplifier has three identical stages each with a time constant of 0.004 S. Calculate the frequency at which the amplifier may oscillate.

3.9 For the feedback amplifier whose Bode plot is given in Fig. Q.3 calculate (a) β, (b) βA at low and medium frequencies.

3.10

(a) Draw the Nyquist diagram of a n.f.b. amplifier that has (i) a gain margin of 3 dB, (ii) a phase margin of 40°.

(b) Draw the Bode amplitude and phase diagrams of an unstable n.f.b. amplifier.

(c) An amplifier has a gain before feedback of $160 \angle 60°$. Feedback is applied with $\beta = 0.01 \angle 50°$. Determine whether the feedback is positive or negative.

3.11 Calculate the voltage gain of the circuit shown in Fig. Q.4.

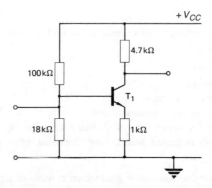

Fig. Q.4

3.12 An amplifier has a gain of 200 and an input resistance of 1200 Ω. Negative feedback is applied to reduce the gain to 50. Calculate the new input resistance. Say which kind of n.f.b. you have assumed.

3.13 An amplifier has a gain of 50 dB that is reduced by n.f.b. to 30 dB. Calculate the change in gain of the amplifier if the gain before feedback falls to 44 dB.

3.14 A three-stage n.f.b. amplifier has three high-frequency time constants, one of which is three times larger than the other two. Calculate the mid-band voltage gain of the amplifier and each of the time constants if the amplifier oscillates at a frequency of 300 kHz, $\beta = 1/120$.

Exercises 4

4.1 The specification of a 741 op-amp includes the following: internal frequency compensation, short-circuit protection, offset voltage null capability, no latch-up. Explain the meaning and importance of each term.

4.2 Fig. Q.5 show the open-loop gain of an op-amp. The amplifier is to be used to provide a closed loop gain of 2000. Determine its bandwidth.

The amplifier is now compensated with a break frequency of 100 Hz. Draw the curve of the compensated gain and determine the bandwidth of the amplifier.

4.3 An op-amp has the following data: input resistance 50 kΩ, input capacitance 100 pF, output resistance 150 Ω.

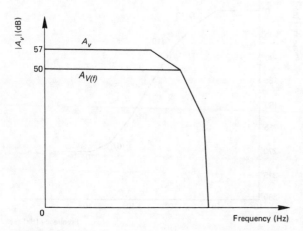

Fig. Q.3

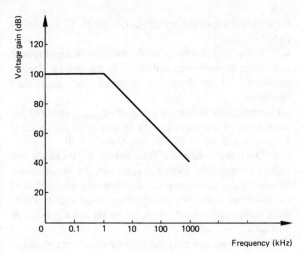

Fig. Q.5

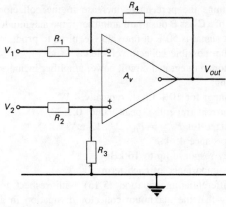

Fig. Q.6

The voltage gain is 6000 into a 1 kΩ load with break frequencies at 2 MHz, 4 MHz and 30 MHz. Feedback is applied so that the overall gain is −100 into a 1 kΩ load. Calculate (a) the required feedback components, and (b) the input and output resistances.

4.4 A 741 op-amp is to be used as an inverting amplifier with a voltage gain of 20 dB and break frequency at 800 Hz. The input impedance is to be 47 kΩ. Calculate suitable component values. Minimize offset output voltage.

4.5 The circuit given in Fig. Q.6 has $R_1 = R_2 = 10$ kΩ, $R_3 = R_4 = 22$ kΩ. Calculate (a) the output voltage when $V_1 = 200$ mV and $V_2 = 500$ mV, (b) the c.m.r.r.

4.6 Show that the output resistance $R_{out(f)}$ of an op-amp connected to give a non-inverting gain is approximately

$$R_{out(f)} = R_{out}R_2/A_vR_1$$

where R_{out} is the output resistance of the op-amp and A_v is the open-loop gain.

4.7 Derive an expression for the inverting voltage gain of an op-amp. Show that this expression will reduce to $-R_2/R_1$ if the open-loop gain is very large. Calculate the voltage gain of an op-amp if $A_v = 3 \times 10^4$, $R_1 = 12$ kΩ and $R_2 = 100$ kΩ.

4.8 Why is it often necessary to connect frequency compensation components to an op-amp? Why is it not necessary for a 741?

An op-amp has $A_v = 10^5$ at low frequencies falling to unity at 1 MHz with a slope of −20 dB/decade. (a) Is this circuit internally compensated? (b) Calculate its break frequency. (c) What will be the break frequency if n.f.b. reduces the gain to 100?

4.9 Define slew rate and explain its importance.

An op-amp has a compensating capacitor of 20 pF into

which a maximum current of 25 µA can be delivered. Calculate the slew rate of the amplifier. For this slew rate calculate the maximum peak value of a sinusoidal output voltage at (a) 1 MHz and (b) 3 MHz.

4.10 List the important parameters for an operational amplifier and discuss how a practical amplifier falls short of the 'ideal'. Explain the concept of a 'virtual earth' in an operational amplifier in the inverting connection.

An uncompensated op-amp has $A_v = 90$ dB and break frequencies of 100 kHz and 1 MHz. Determine the smallest closed-loop gain possible without instability.

Exercises 5

5.1 Draw the circuit of a Class B push-pull amplifier using complementary output transistors. Explain its operation. If the peak value of the collector current of each transistor is 1.6 A, calculate (a) the output power in the 12 Ω load, and (b) the collector efficiency. Take $V_{CC} = 30$ V.

5.2 Draw waveforms of the currents and voltages at various points in a Class B push-pull amplifier. Derive an expression for the efficiency of such a stage.

Calculate the power output and the efficiency of a Class B amplifier in which the peak collector current is 0.7 A and each transistor works into a load of 15 Ω and the supply voltage is 30 V.

5.3 Prove that the collector dissipation in a Class B push-pull amplifier is at its maximum value when the input sinusoidal signal voltage is $2/\pi$ times the maximum possible value. A Class B amplifier uses transistors whose collector dissipation power rating is 1.8 W. If the supply voltage is 16 V determine the maximum power output.

5.4 Determine the percentage increase in the collector dissipation of a Class B push-pull amplifier if the magnitude of the input signal is 50% of the value required to produce the maximum possible collector efficiency.

5.5 Data for an integrated-circuit power amplifier includes the following.

Power output for 10% t.h.d. typically 4 W
Output current (repetitive peak value) 0.9 A
Input voltage for $P_{out} = P_{out(max)}$ 12 mV
Input impedance 45 kΩ
Frequency response up to 16 kHz
Explain the importance of each term.

5.6 By differentiating equation (5.15) with respect to $I_{C(max)}$ show that the maximum collector dissipation in a Class B push-pull amplifier occurs when the peak collector current of each transistor is 0.64 times the maximum permitted peak current.

5.7 For a Class B push-pull complementary-pair amplifier briefly explain (a) how crossover distortion is minimized, (b) why bootstrapping is necessary, and (c) why n.f.b. is applied. Explain briefly why negative feedback has very little effect upon crossover distortion.

A Class B complementary-pair push-pull amplifier is to deliver 10 W to a 8 Ω load. Determine the peak voltage across the load and the peak load current.

5.8
(a) The ratings of the transistors used in a Class B push-pull circuit are $V_{CE(max)} = 20$ V, $I_{C(max)} = 1$ A. Calculate the maximum power that can be delivered to a 4 Ω load.
(b) Two transistors, each having a maximum power dissipation of 1 W, are used in a Class B push-pull amplifier. Calculate the maximum possible output power. What is then the collector efficiency?

5.9 Show that the maximum efficiency of a Class B push-pull amplifier is 78.5%. Also show that if the input signal voltage is reduced by 36% from the value for maximum efficiency the transistor dissipation will increase by 48%.

5.10 Two transistors, each having a rated maximum power dissipation of 500 mW, are used in a Class B push-pull amplifier. Calculate the maximum output power if the efficiency is (a) 75%, (b) 50%.

Exercises 6

6.1 Draw the electrical equivalent circuit for a piezo-electric crystal and explain why it will resonate at two different frequencies.

A crystal is mounted in a holder which has a capacitance of 6 pF. Calculate the two resonant frequencies of the crystal

if the parameters of the crystal are $L = 10$ H, $C_1 = 0.06$ pF and $R = 3000$ Ω.

6.2 Derive expressions for (a) the minimum value of mutual inductance to maintain oscillations in an op-amp Hartley oscillator, (b) the frequency at which the circuit then oscillates.

Determine the minimum value of $A_{v(f)(min)}$ required for oscillations to take place and the frequency of oscillation if $L_1 = 220$ μH, $L_2 = 20$ μH, and $C = 100$ pF.

6.3 Draw the circuit of a Wien bridge RC oscillator and explain its operation. Derive an expression for the minimum voltage gain of the op-amp and the frequency of oscillation. Calculate the values of these quantities if the feedback network components are $R_3 = R_4 = 10$ kΩ and $C_3 = C_2 = 0.01$ μF.

6.4 Use the generalized three-impedance concept to derive expressions for the frequency of oscillation and the maintenance condition of an oscillator.

A Colpitts oscillator is to operate at 4 MHz using an inductor of 80 μH and a transistor whose minimum h_{fe} is 50. Calculate suitable values for the tuned-circuit capacitors.

6.5 Draw and explain the operation of a crystal oscillator.

In a particular circuit the crystal has a shunt capacitance of 12 pF and a series capacitance of 0.06 pF. The crystal is connected to a FET whose input capacitance is 20 pF. Calculate the percentage change in the oscillation frequency when a capacitor of 10 pF is connected across the crystal.

6.6 A crystal has $C_1 = 0.05$ pF and $C_2 = 4$ pF. Express the parallel resonant frequency of the crystal as a percentage of its series resonant frequency.

6.7 A piezo-electric crystal has $L = 50$ mH, $R = 10$ kΩ and $C_1 = 0.02$ pF. Calculate its series resonant frequency.

6.8 Draw the circuit of an op-amp RC oscillator and show that the op-amp must provide a voltage gain of at least 29. Determine suitable component values if the circuit is to oscillate at 10 kHz.

Exercises 7

7.1 An op-amp astable multivibrator has saturated output voltages of ± 12 V. Calculate its frequency of oscillation if $R_1 = 100$ kΩ, $R_2 = 33$ kΩ, $R_3 = 47$ kΩ and $C_1 = 0.01$ μF. Repeat the calculation if the saturated output voltages are +12 V and −10 V. Derive any expressions used.

7.2 Draw the circuit of an IC Schmitt trigger connected to operate as an astable multivibrator and show that the periodic time of its output waveform is given by

$$T = 2C_1R_1 \log_e[V_{DD}/(V_{DD} - V_1)].$$

7.3 Describe the operation of a Schmitt trigger circuit and draw the operating characteristics.

An op-amp of voltage gain 10^4 is connected as a Schmitt trigger with feedback resistors $R_1 = 100\ \Omega$ and $R_2 = 10\ k\Omega$. The output saturation voltages are $V_{o(sat)}^+ = 12$ V and $V_{o(sat)}^-$ $= -11$ V. Determine the approximate voltage levels at which the input signal causes switching to be initiated.

7.4 Figure Q.7 shows the circuit of a sawtooth generator. (*a*) What kind is it? (*b*) Explain the operation of the circuit with the aid of the appropriate waveform diagrams. (*c*) Explain the effect of using a FET as T_2.

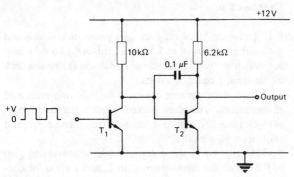

Fig. Q.7

7.5 Draw the circuit diagram of a sawtooth generator using an op-amp and explain the operation of the circuit. Calculate component values to give an output ramp voltage that changes by 10 V/ms. Derive any expression used.

7.6 Draw a circuit diagram to show how a 555 timer IC can be used as an astable multivibrator. Derive an expression for the periodic time of the output waveform. Hence determine suitable values for the external components if the periodic time is to be 1 ms and the waveform is (*a*) approximately square, (*b*) with a mark/space ratio of 1:2.

7.7 Plot to scale the waveform of the voltage across a coil of inductance 6 mH and resistance 10 Ω if the current flowing in the coil is a ramp of peak-peak value 200 mA and the ramp frequency is 12 kHz with 5% of the time used for flyback.

7.8 Figure Q.8 shows the pin connections of a 4001 quad 2-input NOR gate IC. Show how it can be connected to produce a bistable flip-flop.

7.9 A 741 op-amp (slew rate 0.5 V/μs) is to be used as a ramp generator. If a 20 V ramp amplitude is wanted, determine the maximum possible frequency of operation.

7.10 Design an astable multivibrator using a 555 timer to have a mark/space ratio of 1.5:1 at a frequency of 2000 Hz.

7.11 A non-inverting op-amp Schmidtt trigger circuit has $R_1 = 1\ k\Omega$, $R_2 = 50\ k\Omega$, $V_{o(sat)} = \pm 10$ V and $V_R = 1.5$ V. Draw the output waveform of the circuit when the input voltage is a triangular wave of peak value ± 5 V and periodic time 2 seconds.

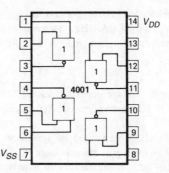

Fig. Q.8

7.12 An op-amp astable multivibrator has $V_{o(sat)}^+ = 14$ V, $V_{o(sat)}^- = -12$ V. The component values are $R_1 = 22\ k\Omega$, $R_2 = 47\ k\Omega$, $R_3 = 100\ k\Omega$ and $C = 0.1\ \mu$F. Calculate (*a*) the frequency of oscillation and (*b*) the mark/space ratio.

7.13 Draw the circuit of a Miller time base and explain its operation. Calculate the amplifier gain necessary for the sweep speed error to be 0.5% with an initial rate of 10 kV/s. The input voltage is a square pulse of peak value 100 V and time duration 10 ms.

7.14 State the amplifier gain required for good sweep linearity in (*a*) a Miller, (*b*) a bootstrap ramp generator.

For either circuit, calculate the gain needed to ensure that the departure from linearity is not more than 1% in a 20 ms sweep. The components affecting the ramp are $C = 0.1\ \mu$F and $R = 1\ M\Omega$.

Exercises 8

8.1 Show that the output signal-to-noise ratio of an amplifier is less than the input signal-to-noise ratio by the ratio $T_s/(T_n + T_s)$ where T_s is the noise temperature of the source and T_n is the noise temperature of the amplifier.

8.2 List the various sources of noise that are generated within an amplifier. State how each of these sources vary with frequency. Explain with the aid of power-density spectrum graphs the meanings of the terms white noise and pink noise.

8.3 An amplifier has a bandwidth of 1 MHz and a noise figure of 10 dB measured with the source at 290 K. A source of noise temperature 580 K is applied to the input terminals of the amplifier when the output signal-to-noise ratio is found to be 20 dB. Calculate the input signal-to-noise ratio of the amplifier.

8.4

(*a*) A bipolar transistor has noise resistances $R_{nv} = 400\ \Omega$ and $R_{ni} = 300\ k\Omega$. Determine the values of the equivalent noise generators v_n and i_n in units of V/$\sqrt{\text{Hz}}$ and A/$\sqrt{\text{Hz}}$, respectively.

(b) A bipolar transistor has noise generators $v_n = 3$ nV/$\sqrt{\text{Hz}}$ and $i_n = 0.8$ pA/$\sqrt{\text{Hz}}$. Determine its equivalent resistances R_{nv} and R_{ni} in a 1 MHz bandwidth.

8.5

(a) Determine the r.m.s. voltage generated in a 1 MHz bandwidth by two 10 kΩ resistors connected (i) in series, (ii) in parallel.

(b) The r.m.s. noise voltage generated in a resistor is 12 μV. What will be the r.m.s. noise voltage if (i) the temperature is doubled, (ii) the bandwidth is halved, (iii) (i) and (ii) occur together?

8.6

(a) The signal-to-noise ratio at the output of an amplifier is 10 dB. Discuss briefly whether this indicates an amplifier with a good or a poor noise performance.

(b) An amplifier has a bandwidth of 200 kHz and a voltage gain of 100. Calculate its noise output voltage due to a 100 kΩ resistor at temperature T_0 connected across the input terminals. Is this the total noise output voltage?

8.7

(a) The input noise to an amplifier is -100 dBm. What power gain is required to give an output signal power of 10 mW if the input signal-to-noise ratio is 40 dB?

(b) An amplifier has a noise figure of 4.5 dB referred to 290 K. What will be the output signal-to-noise ratio if the input signal-to-noise ratio is 36 dB and the source temperature is (i) 290 K, (ii) 580 K?

(c) Calculate the noise output from a circuit which has a noise temperature of 300 K, a power gain of 100, and a bandwidth of 2 MHz if the source temperature is (i) 290 K, (ii) 580 K?

8.8 An amplifier has three stages, the power gains and noise figures of which are given in Table Q.2. Calculate in dB the overall gain and the overall noise figure of the amplifier.

Table Q.2

	Power gain (dB)	Noise figure (dB)
First stage	6	3
Second stage	20	8
Third stage	20	5

8.9 An amplifier has a noise figure of 10 dB and a gain of 30 dB. It is preceded by a pre-amplifier of 4 dB noise figure and 12 dB gain. Calculate the output noise of the

system when the input noise temperature is (a) 290 K, and (b) 200 K. The bandwidth is 1 MHz.

8.10 Prove that the noise figure of an attenuator is equal to its attenuation.

8.11 Show that the output signal-to-noise ratio of an amplifier can be written in the form $T_S/(T_A + T_S) \times$ input signal-to-noise ratio, where T_A = noise temperature of amplifier and T_S = noise temperature of source.

Exercises 9

9.1 Draw the block diagram of a phase-locked loop and outline the function of each block. Explain (a) how the loop acquires lock with an incoming signal and (b) retains lock as the input frequency varies.

9.2 Explain the meaning of each of the following terms used in conjunction with phase-locked loops: (a) capture range, (b) lock range, (c) pull-in time, (d) static phase error, and (e) loop compensation.

9.3 A p.l.l. has a phase detector with a conversion gain of 0.6 V/rad, an amplifier of gain 2, and a v.c.o. of conversion gain 5 kHz/V and a free-running frequency of 75 kHz. Calculate (a) the static phase error when the loop has locked to an input signal of frequency 80 kHz, (b) the lock range.

9.4 A p.l.l. has a loop gain of 20 kHz/rad and uses a low-pass filter with a cut-off frequency of 10 krad/s. Draw its Bode diagram and estimate the phase margin.

9.5

(a) Explain the advantages of using an integrated p.l.l. instead of one with discrete components.

(b) Explain how a p.l.l. can be used to demodulate a frequency-modulated signal.

9.6 Figure Q.9 shows how the frequency of an oscillator can be phase-locked to a high-stability reference oscillator. Explain the operation of the circuit, supporting the explanation with appropriate analysis.

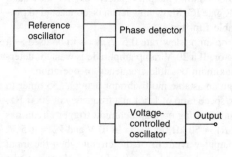

Fig. Q.9

9.7 In the design of a p.l.l. there are two main parameters to be considered: (*a*) the loop gain and (*b*) the natural frequency or bandwidth. Explain the importance of each of these.

9.8 A v.c.o. has a free-running frequency of 50 kHz. When a 0.6 V control voltage is applied the frequency changes to 40 kHz. Calculate the conversion gain of the v.c.o.

Exercises 10

10.1 Discuss the advantages of using an active filter instead of an *LC* design. State three different ways of realizing an active filter and compare their relative merits.

10.2 Draw the circuits of (*a*) a first-order, and (*b*) a second-order filter. How can (*c*) a third-order and (*d*) a fourth-order filter be obtained? Compare two different ways in which a band-pass filter can be made.

10.3 Explain, with supporting equations and analysis, the differences between the Butterworth and the Tchebysheff methods of active filter design. How is the bandwidth of each filter specified?

10.4 An active Butterworth filter is to have a loss of 60 dB at a frequency twice the cut-off frequency. Determine the order of filter necessary. What order filter would be required if a Tchebysheff design were to be used instead? Comment on your result.

10.5 Use equation (10.15) to derive an expression for the transfer function of the high-pass active filter shown in Fig. 10.13.

10.6 Design a second-order Tchebysheff low-pass filter to have 1 dB ripple and a ripple bandwidth of 6 kHz.

10.7 Design a fourth-order Tchebysheff high-pass filter to have 0.5 dB ripple and a ripple bandwidth of 10 kHz.

10.8 Design a fourth-order low-pass Butterworth filter to have a cut-off frequency of 6.5 kHz.

10.9 Design a third-order high-pass Butterworth filter to have a cut-off frequency of 3400 Hz.

Exercises 11

11.1 An average rectified voltage of 35 V across a 20 Ω load is obtained from a 240 V 50 Hz mains supply. The transformer and bridge rectifier used can be considered to be ideal components. Draw the circuit diagram and calculate (*a*) the turns ratio of the transformer, and (*b*) the p.i.v. of each diode. A 1000 μF capacitor is now connected in parallel with the load resistor; calculate the d.c. voltage across the load.

11.2 A power supply connected to a 20 V 50 Hz supply provides 30 V d.c. across a 4000 Ω load. The load has a 8 μF capacitor connected in parallel with it. Calculate (*a*) the turns of ratio of the transformer, (*b*) the output ripple voltage, and (*c*) the ripple factor.

11.3 A transformer with a 25 V secondary winding is used with a bridge rectifier to supply 2 mA to a 10 kΩ load. Calculate (*a*) the load voltage, (*b*) the ripple voltage, and (*c*) the ripple factor. The value of smoothing capacitor is 22 μF.

11.4 A full-wave rectifier unit uses two diodes whose resistances when conducting are constant at 30 Ω. The unit supplies a d.c. current of 30 mA to a 200 Ω load. Calculate (*a*) the r.m.s. transformer voltage, (*b*) the peak current in the diodes, (*c*) the p.i.v. of the diodes, (*d*) the power dissipated in each diode, and (*e*) the a.c. input power if the transformer is 90% efficient.

11.5 Design a full-wave rectifier power supply with capacitor smoothing that can supply a current of 20 mA at 20 V to a load with less than 0.5 V peak−peak ripple.

11.6 Describe the operation of a voltage regulator circuit. List the advantages of using an integrated-circuit regulator. Show how the following features are provided in a power supply: (*a*) remote sensing of the load voltage, (*b*) crow-bar over-voltage protection, and (*c*) fold-back over-current protection.

11.7 Explain the difference between load and line regulation of a power supply. Explain, with the aid of a block diagram, the action of a series voltage regulator. Such a circuit has the following data: gain of voltage comparator = −160, gain of series regulating transistor = −20, output resistance of rectifier feeding regulator = 80 Ω. Calculate the change in the output voltage that occurs if the load current changes by 10 mA.

11.8 The main disadvantage of the linear voltage regulator is the power that is dissipated in the series regulating element. Explain how the use of a switched-mode regulator overcomes this disadvantage. Draw the basic diagrams for the flyback and forward converter s.m.p.s. and briefly explain their action. Which of them would be used for a low-power application?

11.9 In an s.m.p.s. there is no transformer reaction to limit the current that flows when the power supply is first switched on. Suggest a method by which the current drawn as the smoothing capacitor charges up can be limited to a safe level. Explain why a s.m.p.s. is a source of electrical noise and suggest some ways of reducing this noise.

Answers to Numerical Exercises

1.1 31 °C/W

1.2 5.43 °C/W, 10.77 W

1.3 575, (a) −10.3%, (b) −4.35%

1.4 (a) 2 kΩ, 20 kΩ, (b) 140

1.5 (a) 44.72, (b) 19.61, (c) 9.95

1.6 6.8 MHz

1.7 1947 Ω

1.9 2.34 °C/W

1.10 g_m = 164 mS, r_{ce} = 12.5 kΩ, $r_{b'e}$ = 1000 Ω, $r_{bb'}$ = 100 Ω

1.11 6 ∠ −37°

1.12 (a) 5, 10, (b) 7, 96 pF

1.13 9800, 37.3 kΩ, 262.7

2.1 (a) 114.59 × 10⁶, (b) 2.865 MHz

2.2 0.37

2.3 (a) 331 Ω, (b) 191, (c) 8.325 MHz

2.4 0.47 μF, 0.447

2.5 (a) 275.74, (b) 4800 Ω, (c) 230, (d) 3077 Ω

2.6 119 pF, 9.5 pF

2.7 156.81 kΩ

2.8 417 Ω

2.11 (a) 26 dB, (b) 20°, (c) 540 Hz, (d) 2 kHz, (e) 2 kHz

2.12 360 pF, 68.2, 9.5 kHz

3.4 0.0197, 3333.3

3.5 19

3.6 (a) 82.14, (b) 115.46 ∠ −150°

3.7 (a) 0.4 dB, (b) 1.308 MΩ

3.8 23 Hz

3.9 1.75 × 10⁻³, 1.24

3.10 negative

3.11 4.7

3.12 4.8 kΩ

3.13 0.834 dB

3.15 707, 0.685 μs, 0.685 μs, 2.055 μs

4.3 (a) 1 kΩ, 100 kΩ, (b) 1017 Ω, 2.525 Ω

4.5 0.66 V, zero

4.7 8.33

4.8 (a) yes, (b) 10 Hz, (c) 1000 Hz

4.9 1.25 V/μs, (a) 0.2 V, (b) 0.066 V

4.10 70 dB

5.1 (a) 7.68 W, (b) 50.27%

5.2 1.84 W, 55%

5.3 9 W

5.4 41.28%

5.7 25.3 V, 3.16 A

5.8 (a) 10 W, (b) 5 W, 71.4%

5.10 (a) 3 W, (b) 1 W

6.1 205.47 kHz, 206.49 kHz

6.2 11, 1.027 MHz

6.3 3, 1592 Hz

6.4 20.2 pF, 1010 pF

6.5 0.022%

6.6 100.625%

6.7 5.033 MHz

7.1 371 Hz, 370 Hz

7.3 0.1188 V, −0.1089 V

7.5 If V_{CC} = 15 V, 100 kΩ, 15 μF

7.9 3979 Hz

7.12 137 Hz, 0.93:1

7.13 99

7.14 9

8.3 27.4 dB

8.4 (a) 2.53 × 10⁻⁹ V/√Hz, 2.31 × 10⁻¹³ A/√Hz, (b) 562 Ω, 25 kΩ

8.5 (a) (i) 17.89 μV, (ii) 8.95 μV, (b) (i) 16.97 μV, (ii) 8.485 μV, (iii) 12 μV

8.6 (b) 1.79 mV, No

8.7 (a) 70 dB, (b)(i) 31.5 dB, (ii) 33.19 dB, (c)(i) 1.63 × 10⁻¹² W, (ii) 2.43 × 10⁻¹² W

8.8 46 dB, 5.23 dB

8.9 (a) 132 pW, (b) 163 pW

9.3 47.8°, 18.85 kHz
9.8 16.67 kHz/V

10.4 10, 7

11.1 (*a*) 6.17:1, (*b*) 55 V, 45 V

11.2 (*a*) 9.8:1, (*b*) 4.69 V, (*c*) 9%
11.3 (*a*) 20 V, (*b*) 0.91 V, (*c*) 4.55%
11.4 (*a*) 7.64 V, (*b*) 47 mA, (*c*) 21.6 V, (*d*) 51 mW, (*e*) 282 mW
11.7 0.248 mV

Index

A.C. current gain, of a bipolar
  transistor 3
Active filter 239
  Sallen and Key 240
    high-pass 244
    low-pass 241
  state variable 246
Advantages, of n.f.b. 86
Amplifier,
  audio-frequency large-signal 131
    discrete component — Class A 131
      collector dissipation 132
      collector efficiency 131
      harmonic distortion 133
      output power 133
    discrete component — Class B 134
      bootstrapped 137
      collector dissipation 135
      collector efficiency 135
      cross-over distortion 134
      design of 138
      harmonic distortion 138
      negative feedback 136
      power MOSFET 144
      integrated circuit 140
      LM 380 data sheet 141
    audio-frequency small-signal 39
      bandwidth 66
      bias circuit 39–40
      stability function 41
      design of 44
  differential amplifier 70
  emitter follower 49
    bootstrapped 51
    dual power supply 50
  equivalent circuit 46
  Miller effect 62
  multiple stages 51, 66
  negative feed back, *see* negative
    feedback
  op-amps, *see* op-amps
  undecoupled emitter resistor 47
  voltage gain
    high-frequencies 61, 64
    low-frequencies 57, 59
    mid-frequencies 45
  integrated circuit 51
    LM 381 data sheet 53

tuned radio-frequency 66
  bandwidth 68
  coupling methods 67
Amplitude-frequency distortion 87
Astable multivibrator
  555 timer 178
  IC 176
  NAND/NOR gate 177
  op-amp 176
Available noise power 197
Available power gain 203

Bandwidth
  a.f. amplifier 66
  n.f.b. amplifier 88
  op-amp 105, 107
  r.f. amplifier 68
Barkhausen criterion 146
Bias circuit, bipolar transistor 40
Bias current, of op-amp 100
Bipolar transistor
  bias 39
    choice of operating point 40
    design of 44
    stability function 41
  capacitances 12
  collector leakage current 2
  current gain,
    a.c. 3, 19, 26
    d.c. 2
  data sheets, 1, 31, 32, 38
  equivalent circuit 17
    h parameter 18
    hybrid π 22
    y parameter 27
  gain-bandwidth product 31
  hybrid parameters 18
  hybrid π parameters 22
  junction temperature 5
  maximum collector current 9
  maximum collector voltage 10
  Miller effect 24
  power dissipation 5
  thermal capacitance 8
  thermal resistance 5
Bistable multivibrator 162
  op-amp 163

Bode plot,
  a.f. amplifier 59, 61
  n.f.b. amplifier 93
  phase-locked loop 216, 221, 223
Bootstrapping,
  emitter follower 51
  power amplifier 137
  ramp generator 185
  voltage follower 121
Burst noise 198
Butterworth filter 229, 233, 245

Capacitances,
  bipolar transistor 12, 23
  FET 16
Clamping and clipping, of
  waveforms 189
Class A power amplifier 131
Class B power amplifier 134
Collector leakage current 2
Colpitts oscillator 153, 154
Common-mode gain 71
Common-mode rejection ratio (c.m.r.r.)
  71, 108
Compensation,
  of n.f.b. amplifier 94
  of op-amp 122
  of p.l.l. 221
Coupling methods
  in a.f. amplifiers 51
  in tuned r.f. amplifiers 67
Cross-over distortion 134
Crowbar circuit 259
Crystal oscillator 159
Crystal, piezo-electric 157
Current gain, of bipolar transistor 2, 3,
  19
Current limiting 259
Current mirror 69
Current-series n.f.b. 74, 80
Current-shunt n.f.b. 74, 82
Current source 68
Cut-off frequency, of bipolar transistor
  3

Darlington connection 12, 14, 139
Data sheets
  a.f. power amplifier 140
  a.f. pre-amplifier 53
  bipolar transistor 1, 31, 32, 38
  op-amp 109, 110, 112, 116
D.C. stabilization, of bias circuit 40
Design of
  bias circuit 44
  Class B push-pull amplifier 138
  Colpitts and Hartley oscillators 156
  op-amp circuit 128
  RC oscillator 147
  Wien bridge oscillator 150
Differential amplifier
  bipolar transistor 70
  op-amp 126
Differential gain 71

Distortion
  amplitude-frequency 87
  crossover 134
  intermodulation 89
  non-linearity (harmonic) 89,
    133, 138

Electro-magnetic interference (e.m.i.)
  194
Emitter follower 49
  bootstrapped 51
  dual power supply 50

Field effect transistor (FET) 14
  capacitance 17
  equivalent circuit 30
  gain-bandwidth product 31
  maximum drain-source voltage 16
  mutual conductance 15
  power dissipation 17
  power MOSFET 17
  temperature effects 15
Filter
  active 239
  Butterworth 239, 243, 245
  first-order 231
  gyrator 248
  high-pass 244
  low-pass 215, 231, 233, 236, 241
  Sallen and Key 240
  second-order 232
  state variable 246
  Tchebysheff 230, 236, 246
  555 timer 161
  565 phase-locked loop 224
Flicker noise 197
Follower, emitter 49
  voltage 121
Frequency compensation
  of n.f.b. amplifier 94, 97
  of op-amp 122
Frequency multiplication 227
Frequency stability, of oscillator 156
Full-wave rectifier 249

Gain-bandwidth product 27, 31, 106
Gain margin, of n.f.b. amplifier 92
Gyrator 248

Harmonic distortion 89, 133, 138
Hartley oscillator 153, 155
h parameters, of bipolar transistor
  2, 3, 18
  calculation of current gain 19
    of input resistance 20
    of output resistance 22
    of voltage gain 21
Hybrid π equivalent circuit 22
  calculation of current gain 26
    of input admittance 24
    of voltage gain 23, 25

Input bias current, of op-amp 100
Input offset current, of op-amp 101

Input offset voltage, of op-amp 102
Instrumentation amplifier 127
Integrated
  a.f. power amplifier 140
  a.f. pre-amplifier 53
  multivibrator 172
  op-amp, 101/201/301 112
    741 110
    various 114
  phase-locked loop 224
  Schmidtt trigger 167
  timer 161
Interference 193
  reduction of 195
Intermodulation distortion 89

Junction temperature 5

LC oscillator 151
Linear voltage regulator 253
LM 380 a.f. power amplifier 140
LM 381 a.f. pre-amplifier 52
Low pass filter 215, 231, 233, 234, 241

Measurement of
  noise figure 209
  noise temperature 210
Miller effect 24, 62, 183
Miller ramp generator 182
Monostable multivibrator 168
Multiple stages, in amplifiers 51, 66
Multivibrators
  astable 174
    555 timer 178
    IC 176
    NAND/NOR gate 177
    op-amp 174
  bistable 162
    op-amp 163
  monostable 168
    555 timer 171
    IC 172
    NAND/NOR gate 169
    op-amp 168
    Schmidtt trigger 170
Mutual conductance, of bipolar
  transistor 4, 15

Negative feedback 40, 49, 56, 73, 136
  advantages of 86
  current-series 74, 80
  current-shunt 74, 82
  distortion 87, 89
  gain stability 86
  stability of 89
    Bode diagram 93
    gain margin 92
    log-gain plot 96
    Nyquist diagram 90
    phase-lead compensation 94, 97
    phase margin 93
  voltage-series 74, 75
  voltage-shunt 74, 84

Noise 192
  available power 198
  burst 198
  corner frequency 199
  equivalent resistance 199
  flicker 197
  op-amp 199
  pink 193
  popcorn 198
  power density spectrum 192
  reduction of 195
  shot 197
  sources of 193
  white 192
Noise factor (or figure) 203
  of cascaded networks 207
  effective 207
  measurement of 209
  spot 206
  variation with frequency 205
Noise temperature 208
  of cascaded networks 209
  measurement of 210
Non-linearity distortion 89, 133, 138
Nyquist diagram, of n.f.b. amplifier 90

Operational amplifiers (op-amps) 99
  bandwidth 105
  common-mode rejection ratio
    (c.m.r.r.) 108
  data sheets
    101/201/301 112
    741 110
    various 114
  design of circuits 128
  differential amplifier 126
  drift 104
  frequency compensation 122
  full-power bandwidth 107
  impedance of 100
  input bias current 100
  input offset current 101
  input offset voltage 102
  instrumentation amplifier 127
  internal circuitry 109
  inverting amplifier 99, 114
  noise 199
  non-inverting amplifier 99, 119
  selection of 124
  slew rate 107
  transient response risetime 106
  unity-gain bandwidth 106
  voltage follower 121
Oscillators 145
  Colpitts 153, 154
  crystal 159
  frequency stability 156
  Hartley 153, 155
  LC 151
  principle 146
  RC 146
  tuned collector 151
  voltage-controlled 214
  Wien bridge 148

Phase detector 211
Phase-locked loop (p.l.l.) 211
  analysis 218
  applications of 227
  capture range 218
  dynamic behaviour 219
  integrated (565) 224
  lock range 218
  loop compensation 221
  low-pass filter 215
  phase detector 211
  principle of 216
  v.c.o. 214
Phase margin, of n.f.b. amplifier 93
  of p.l.l. 222
Piezo-electric crystal 157
Pink noise 193
Popcorn noise 198
positive feedback 73, 146
Potential-divider bias circuit 40
Power MOSFET 17
  push-pull amplifier 144
Power supplies 249
  crowbar circuit 260
  current limiting 259
  full-wave rectifier 249
  over-voltage protection 260
  remote sensing 259
  voltage regulation 252
    linear 253
    switched-mode 255

Radio-frequency interference
    (r.f.i.) 194
Ramp generator 182, 185, 188
Ramp waveform 180
RC oscillator 146
Rectifier, full-wave 249
Remote sensing 259

Sallen and Key active filter 240
Sawtooth generator
  Bootstrap 185
  555 timer 188
  Miller 182

Schmidtt trigger 163
  IC 167
  op-amp 165
Shot noise 197
Slew rate, of op-amp 107
SOAR diagram 10
Stability
  of bias circuit 41
  of n.f.b. amplifier 89
  of oscillator 156
  of p.l.l 221
Switched-mode power supply 255
  forward convertor 256
  flyback convertor 258

Tchebysheff filter 229, 236, 246
Temperature effects,
  bipolar transistor 4
  FET 15
Thermal agitation noise 196
Thermal capacitance 8
Thermal resistance 5
Thermistor 149
Timer IC (555) 161
  as astable multivibrator 178
  monostable multivibrator 171
  ramp generator 188
Transition frequency 3
Triangle-wave generator 187
Tuned amplifier 66
Tuned-collector oscillator 151
Voltage-controlled oscillator 214
Voltage follower 121
Voltage regulation, of power
    supply 252
  linear 253
  switched-mode 255
Voltage-series n.f.b. 74, 75
Voltage-shunt n.f.b. 74, 84

White noise 148
Wien bridge oscillator 192

y parameters, of bipolar transistor 27